Facilities Evaluation Handbook:
Safety,
Fire Protection and
Environmental Compliance

FACILITIES EVALUATION HANDBOOK:
SAFETY, FIRE PROTECTION AND ENVIRONMENTAL COMPLIANCE

BY

K.L. PETROCELLY, F.M.A., C.P.E.

Published by
THE FAIRMONT PRESS, INC.
700 Indian Trail
Lilburn, GA 30247

Library of Congress Cataloging-in-Publication Data

Petrocelly, K. L. (Kenneth Lee), 1946 -
  Facilities evaluation handbook : safety, fire protection, and environmental compliance / by Kenneth L. Petrocelly.
    p. cm.
  Includes index.
  ISBN 0-88173-114-5
  1. Buildings--Performance. I. Title.
TH453.P49   1991   690'.22--dc20   91-25530
                                                        CIP

Facilities Evaluation Handbook: Safety, Fire Protection, and Environmental Compliance by Kenneth L. Petrocelly.
©1992 by The Fairmont Press, Inc. All rights reserved. No part of this publication may be reproduced or transmitted in any form or by any means, electronic or mechanical, including photocopy, recording, or any information storage and retrieval system, without permission in writing from the publisher.

Published by The Fairmont Press, Inc.
700 Indian Trail
Lilburn, GA 30247

Printed in the United States of America

10 9 8 7 6 5 4 3 2 1

ISBN 0-88173-114-5   FP

ISBN 0-13-299157-8   PH

While every effort is made to provide dependable information, the publisher, authors, and editors cannot be held responsible for any errors or omissions.

Distributed by Prentice-Hall, Inc.
A Simon & Schuster Company
Englewood Cliffs, NJ 07632

Prentice-Hall International (UK) Limited, London
Prentice-Hall of Australia Pty. Limited, Sydney
Prentice-Hall Canada Inc., Toronto
Prentice-Hall Hispanoamericana, S.A., Mexico
Prentice-Hall of India Private Limited, New Delhi
Prentice-Hall of Japan, Inc., Tokyo
Simon & Schuster Asia Pte. Ltd., Singapore
Editora Prentice-Hall do Brasil, Ltda., Rio de Janeiro

TO ALL ASSESSORS OF FACILITIES—
SHELLBACKS AND POLLIWOGS ALIKE!

Preface

Aside from the owner, no one is more protective of an organization's physical plant or familiar with its character than the facility's manager... and even the owner gets lost when the conversation looms into the nuts and bolts of the operation. Sometimes referred to as the Building Engineer or Maintenance Director, the Facilities Manager is the personal attending physician to the organization's physical self; sensitive to its every belch and whimper. His/her daily observations and probing help maintain the heart of the operation at a rhythmic pulse. But even the fittest of bodies can benefit from an annual physical; and this is what I'm prescribing here. At a minimum, facility evaluations identify present problems for future planning, provide for accurate tracking of labor and material expenditures and act as a base of information for justifying capital, operating and man-hour budget forecasts, from which plans for correction can be formulated.

Contents

Chapter 1   Assessing the Physical Plant 1
the evaluation rationale, a process overview, the need to evaluate the benefits derived, physical and functional elements, the third faction, the inspection perspective

Chapter 2   Preparing for the Process 13
outlining a plan, one man's poison, references and review documents, assembling a team, surveys and forms, gearing up

Chapter 3   The Team Leader's Walk-Through 25
honing your homing skills, the team leader's tool pouch, comparing facilities services, a typical tour... *the powerhouse, the grounds, other mechanicals, facility interiors...*

Chapter 4   Dealing with Risk 37
The risk perspective, the concept of risk, diagnosing exposure levels, managing the threat... *objectives, risk distribution, controlling strategies,* taking corrective action

Chapter 5   Safety Management 49
program characteristics, the safety committee, policies and procedures, department safety rules, new employee orientation, the supervisor's role

Chapter 6   Fire Protection 69
fire codes/regulations, the company fire program, life safety code history, structural containment, life safety systems, storage of flammables

Chapter 7	Electrical Safety	81

elements of electrical shock, program considerations, maintenance of high-voltage substations, care of panelboards, systems maintenance, ground fault interruption, single phase protection, emergency power generation, safety tips

Chapter 8	Environmental Compliance	105

petroleum storage laws, worker right-to-know, asbestos, hazardous waste, clean air regulation, water quality, safe drinking water, pesticides, toxic clean up

Chapter 9	Scoring the Evaluation	125

site characteristics and grounds maintenance, interior and exterior building structure, physical plant systems and equipment, safety and fire protection, environmental compliance, department operations

Chapter 10	The Evaluation Report and Corrective Action Plan	175

laying out the report, introducing and summarizing the evaluation, transcribing the scorecard, correlating the support data, fashioning the corrective action plan

Appendix A	Safety Committee Charter	187
Appendix B	Supervisors' Safety Worksheets	197
Appendix C	Statement of Construction and Fire Protection	205
Appendix D	Glossary	223
Appendix E	Tables	243
Index		269

CHAPTER 1
ASSESSING THE PHYSICAL PLANT

Valhala!... I knew it would come to me if I worried over it long enough... or is it Utopia? Help me out. You know what I'm looking for; the name Facilities Managers use to describe their domains once all the systems and programs have been installed and implemented, ...after all the service agreements have been signed and staffing is at par level. No such place you say? Of course I agree. Why else would they spend good time and money to frequently assess their already fully functioning operation and the structures that house them? Moreover, when is it appropriate for them to do so? And when they do, what do they look for? I'll answer these questions and explain the how-to's of the evaluation process as we proceed.

THE EVALUATION RATIONALE

As I see it, there are three stages in the life of a facility when extraordinary scrutiny of its assets and operations is called for: when construction ends, during its normal operating life and prior to its disposal as an asset. Each calls for a different investigatory perspective and level of examination and can generally be categorized as follows:

Project Close Out
...a point in a building project at which the construction process is considered complete and commissioning of the electrical and mechanical systems takes place. Prior to close out, discrepancies are noted, prompting corrective actions to be taken and initial system performance data are

acquired for future comparisons. Once the change orders and punch-list items have been satisfied, all as-built drawings and equipment specifications (as installed) are turned over to the owner, along with all applicable operating and service manuals. At this stage, inspections fall within the realm of the Project Manager.

Figure 1-1. Equipment is commissioned at project close out, only after all operating parameters are met.

Post Occupancy

...any time in the period after which the physical plant is considered operational and is officially occupied by permanent tenants. Post-occupancy evaluations fall into three main classifications:

1. *indicative*
2. *investigative*
3. *diagnostic*

An *indicative* POE is comprised of a simple walk-through, referencing as-built drawings, specifications and standard building checklists to gain a down and dirty idea of a facilities overall condition.

An in*vestigative* POE entails the more involved processes of research, testing and base data comparisons. It takes longer and costs more to conduct than an indicative POE but is less comprehensive than a diagnostic POE.

A *diagnostic* POE is the most complicated and time-consuming of the three methods, utilizing sophisticated investigatory and analytical tools, actuarial and statistical data.

Which method you choose for assessing your operations depends on the depth of the information you require, how much time you can devote to the project and the resources you have at your disposal to conduct it. Post-occupancy evaluations can be scheduled or performed as needed under the auspices of the Facilities Manager.

Pre-Disposition

Whether companies are selling out, moving to greener pastures or simply liquifying a few of their assets, from time to time they put certain of their commercial properties (buildings) on the market. An independently derived, pre-sale appraisal, used to compare notes with the buyer, could well be the margin between profit and loss for the owner. Although the Building Manager or Plant Engineer is generally considered the best source for information regarding a facilities operating history and present condition, at this stage, it would behoove an owner to hire a certified appraiser to perform the evaluation; for two reasons:

1) professional appraisers assess commercial properties for a living and are trained in property valuation techniques.

2) conclusions drawn by outsiders are usually considered less biased than those of in-house personnel.

A Process Overview

For purposes of this text, we'll be exploring the post-occupancy aspect of evaluation, dealing with the operational side of the issue; we'll leave the headaches of setting up and moving on to others. As mentioned earlier, there are three basic strategies used when performing a post-occupancy evaluation; a cursory surveillance (indicative), a more intensive probing (investigative) and the most complex (diagnostic).

The *indicative* POE, as its name implies, provides a quick "indication" of the functionality of a building, its systems & equipment, based on superficial inspections of short duration. The facilities as-built drawings and specifications are reviewed by the evaluator or evaluating team prior to a systematic walk-through using standard checklists for recording obvious discrepancies and other items of note. The intent at this level is strictly to observe status and condition; no testing is performed nor is any performance criteria developed. The *investigative* POE calls on the investigatory powers of the evaluator to apply deductive reasoning to test results and base-data comparisons.

Conclusions are drawn as to need from which a viable plan of corrective action is formulated. The *diagnostic* POE is by far the most involved and comprehensive of the three approaches, employing a wide variety of research methods and techniques. This form of the process incorporates observation and analysis, as in the foregoing methods, along with surveys, questionnaires, physical measurements and conceptualizations—taking all aspects of performance and compliance into account.

THE NEED TO EVALUATE

By the nature of their work, Facilities Managers are constantly on the prowl, ever observant and cognizant of goings on throughout their plants. Like mother hens, they faithfully tend to their flocks of window panes, air-compressors and alarm systems. They keep their fingers on the pulses of their operations through the activities and recordings of subordinate personnel, regular tours of public areas and occasional visits to out-of-the-way spaces when trouble occurs. Effectively, as facilities are managed over time, such concerted vigilance is tantamount to performing an ongoing indicative POE. But infrequently, a more diligent scrutiny of the premises is called for. To my mind, there are several times when a more in-depth evaluation may be desirable, for example:

- 6 months after assuming responsibility for an existing operation
- when the facilities base-data is sketchy, questionable or non-existent

Assessing The Physical Plant

- prior to the expiration of major warrantees and guarantees
- as a consequence of changes in regulatory body requirements
- during extended periods of shut down (scheduled and unscheduled)
- when considering large capital outlays for systems upgrades
- after major renovations have been performed

Not every evaluation needs to encompass the whole of the operation. The magnitude of conducting a diagnostic POE makes it too cost-prohibitive and time-consuming to be performed plant-wide on a regular basis; yet its parameters meet the requirements of myriad specific concerns, such as roof surveys and equipment change-outs. In such cases, the diagnostic format can be utilized to resolve problems requiring thorough research and analysis without resorting to a full-blown POE. Much of what the evaluating Facilities Managers do is done just this way, as it affords detailed attention to individual areas as the need arises. But plant-wide, the investigative POE provides the best return for the resources expended.

THE BENEFITS DERIVED

Most plant engineers and building managers will admit that it took more than a few long walks and discussions over coffee to totally familiarize themselves with their facilities. Let's face it; there's just too much to learn and too little time to learn it. Besides orienting themselves to the layout of the building, finding out who is on which floor, tracing out the utility feeds and getting to know their people, they're constantly putting out fires, acting as referees in tenant disputes, arguing with inspectors, and submitting project and budget packages for approval ... among other things. Depending on the size, scope and complexity of their operations, managers may put in 6 months to 2 years at a new location before feeling comfortable there.

So what is THE main benefit of performing an in-depth POE? In my experience, it's the calm that comes over you when you feel like you're finally gaining control of your operations. During the honeymoon period a building manager relies on blind faith, pure luck and the grace of God to pull him through. To an extent, a well thought-out and executed evaluation, enables that manager to master his own destiny.

Some other benefits associated with POE's are:

- identification of present problems for future planning
- access to a master equipment list for monitoring performance, history, care, depreciation and life expectancy
- establishment of adequate, cost-effective parts and materials inventories
- enables phase-out of non-standard components and fixtures
- provides a more accurate track of labor and material expenditures
- acts as a base of information from which plans for correction can be formulated
- aids in the authoring of a comprehensive facilities S.O.P. manual
- provides capital, operating & man-hour budget justifications

Physical and Functional Elements

The inspection of building components and space involves two factions:

1) the physical, which deals with condition, capacity and end product

and

2) the functional, which concerns itself with the practicality of adjacencies and ergonomic constraints.

In most cases each calls for individual consideration, but at times they are difficult to separate, due to the legal & financial mandates that bind them. As I've never aspired to becoming a barrister (though I've had similar title conferred on me) and since my designation reads CPE not CPA, I've decided to treat them as components in common. As such, here are some elements to consider during the visual part of the plant inspection:

Assessing The Physical Plant

Figure 1-2. The appearance of a building's exterior reflects how people perceive its interior operations. (*Courtesy of Weil-McLain*)

Site Characteristics

- topography
- pavement-to-dirt ratio
- drainage ability
- view/noise
- pedestrian access
- vehicular ingress/egress
- micro-climates
- orientation
- loading zones
- handicap access
- easements
- exterior facades
- signage
- adequacy of parking
- lighting
- security
- groundskeeping

Building Characteristics

- foundation stability
- Structural integrity
- load bearing walls
- compartmentalization
- shoring/bracing
- handicap accessibility
- design loading
- roof/wall systems

- doors/windows
- insulation
- functional adjacencies
- interior partitioning
- life safety considerations
- flame spread of finishes
- hazardous material storage
- acoustics

Mechanical Systems

- ample capacity
- fire protection
- component tranes
- smoke control
- circuit zoning
- freeze protection
- energy efficiency
- air supply/exhaust
- water treatment
- space comfort
- duct/cable routes
- emergency power/UPS
- lighting
- communications components
- clocks/time-clocks
- computer networks
- surveillance/keying
- energy management
- elevators
- waste disposal
- preventive maintenance
- parts inventory
- component redundancy
- power factors

THE THIRD FACTION

Though inextricably linked to a facility's existence and condition, an area often overlooked during assessments is Department Operations. Elements of this faction overlap both the physical and the functional; neither of which could or would exist without being subjected to the scrutiny and manipulation of operations and management personnel and processes. In no particular order, here are some items you can add to your list of elements for consideration:

- man-hour, capital and operating budgets
- records retention
- preventive maintenance
- parts/materials inventories
- work requisitioning
- policies/procedures

Assessing The Physical Plant

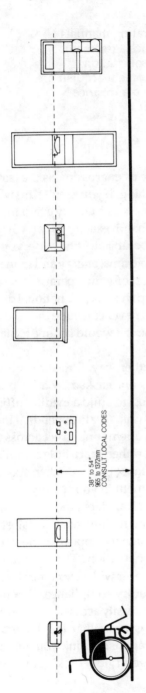

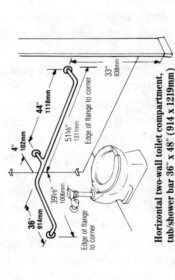

Figure 1-3. Barrier-free design is no longer the option it once was. (*Bobrick Washroom Equipment, Inc.*)

- quality assurance
- systems certification
- rounds/routines
- employee performance
- management plans
- blue print index
- operating manuals
- service agreements
- bid solicitation
- equipment acquisition
- code compliance
- ... etc

THE INSPECTION PERSPECTIVE

Surveying a physical plant can become as engrossing as the last 2 minutes of a Super Bowl game, and as punishing, if you aren't sure why you're there. Before taking on as awesome a task as a facility evaluation, it makes good sense to determine beforehand exactly what you're attempting to find out about what you'll be looking at. Otherwise, you'll end up spinning your wheels, wasting valuable time, energy and money.

If you attempt to scrutinize every aspect of your operations from every vantage point, you're costs could end up rivalling the national debt and like so many government projects, you'd never conclude the assessment, as once you'd reached the last component it would be time to start the process over again.

Still not clear? Let me put it to you another way. (I'm sure you've been posed with a similar quandary). I was once asked by a hospital administrator — "Ken, how much would it cost to build a medical office building?" After a short ponder, my reply was "It all depends. How much does it cost for a new car?" What I was attempting to get across to him wasn't how stupid his question sounded, rather, that a little common sense was called for. So too, common sense must be used when speculating on the status of the buildings & equipment in your charge.

How would you evaluate a steam boiler? It all depends on how you view it. Are you concerned with operating and maintenance costs? The price and amount of fuel it consumes? Then finance appears to be your major concern. What about age and hours of operation? Those could be financial or safety based. Are you polluting the air with stack emissions? Those observations enter the realm of regulatory compliance. Can the control panel be locked out? Is the boiler room, easily accessible from the outside? Is the room well lit? Safety/security issues. What about operating logs and records retention? Legal implications? I think you get the picture. Common sense dictates that you do a good job of looking but that you only look at what you need to.

(Form 1) MAINTENANCE WORK REQUISITION

DATE:	DEPT.:	LOC:	CC #
DESCRIPTION OF WORK NEEDED:			(CHECK ONE)
			MAINT.
			OTHER
DEPARTMENT HEAD SIGNATURE:			

FOR PLANT OPERATIONS DEPT. USE ONLY URG ☐ ROUT ☐ DEF ☐

DATE RECEIVED	APPROVED	JOB ESTIMATE	COMPLETED
ASSIGNED TO	NOT APPROVED	MH / MAT	DATE / BY
ASSIGNED BY	REASON:	MATERIALS USED	TIME SPENT
DATE ASSIGNED			REMARKS:
MAINTENANCE TYPE	ACTUAL COST		
CORRECTIVE	LABOR – $		
PREVENTIVE	MATERIAL – $		
ADM. APPROVAL	TOTAL – $		

PLANT COPY

Figure 1-4. Nothing gets done without one of these.

CHAPTER 2
PREPARING FOR THE PROCESS

If it's true that "... good things come from adversity," then hold onto your hard hat: the format I'm about to share with you (Figure 2-1) resulted from headache, heartache, trauma and pain. It was taken from the Table of Contents of my most recent facilities evaluation report, having evolved into its present form via trial and accomplishment over the years.

OUTLINING A PLAN

Okay. So you've decided just what you're going to be looking for. Now, what? There is no one standard approach to planning a facility evaluation. As already alluded to, they are contingent on several factors, including but not limited to:

- the type of evaluation to be performed
- who is requiring it
- from what perspective observations will be made
- the resources available for its performance
- what archival information is contained in the files
- the size and age of the facility
- ... etc.

Thus, every assessment is as individual as the operations, equipment, people and buildings they're housed in. But, you say, we've got to start somewhere! True. And, just as you'd address any other complex problem in your plant, it's simply a matter of getting your ducks lined up, one behind the other. As singular as assessments are, four common threads run through each of them. Keep this logical progression in mind as you outline your plan, to help you think through the project before its implementation:

- review of drawings, specifications and documents
- facility and site tours
- inspection and analysis of conditions found
- generation of reports and plans for correction

ONE MAN'S POISON ...

Though I've bitten into difficult operations before, they were nothing more than fodder to tide me over until I got to the main course. In the late 1980's I took on a facility that, for lack of a better description, was a veritable feast of misfortune. Brother, if dilapidated structures and debilitated systems are the orphans of calamity, then I found their decrepit mother.

The building's electrical systems were overloaded, its six in-house transformers were contaminated with PCB's, there was no UPS system and its four emergency generators were unreliable at best. The incinerators smoked, four of the four underground oil tanks leaked and of course there were the asbestos problems. Sound pretty bad? Hell, I didn't get to the good parts. Reviewing my daily journal for the period, here are some entries I made during my first two weeks on the job:

- Basement floor severely flooded during a thunderstorm; water backing up through sanitary drains; received several calls from irate department heads; no pumps available; had maintenance and custodial personnel use wet vacs, mops and buckets to dry up

- Pulled as-builts and discussed the problem with an architect who was familiar with the building's plumbing systems (found out that the 15-inch street main is actually only a 12-inch combination sewer line).

- Prior Director absconded with the department S.O.P. manual and several important files

- The fan coil units in a recently renovated area are missing parts and/or are inoperable; there is also a history of frozen coils
- The west bank and #12 elevators are bouncing; many indicator lights missing
- Called to the Board Room by the president to discuss maintenance problems needing *immediate* attention: Stained ceiling tile! Rheostat doesn't work!
- Sat in on grievance meeting with personnel director and business agent (dropped grievance for 4-6 weeks until I evaluate the mechanics' ability and job performance)
- The new 1000-ton chiller arrived at the power house; driver wants to know where it goes! (I wasn't aware one had been ordered!)
- I put the prior scheduled retubing of #2 boiler on hold ($73,000) until I could determine if it was justified. (The power house engineer said it was being retubed because —IT WAS TIME !!!
- Discussed prior night's "break-ins" with the security manager.
- Note: there is an apparent lack of standardization re: plumbing/light fixtures, floor/ceiling tile, ... etc.
- The power house supervisor will be out on disability for 6 to 8 weeks with a broken leg suffered in a motorcycle accident
- Received several "too hot" calls. #2 centrivac (still out?) for overhaul (no a/c service agreement)
- Frequent trouble lights on fire warning system annunciator panel causing false alarms and dispatching of trucks by the fire department
- Both incinerators down again! (When were they down originally?)
- Boiler stacks smoking—engineer says it's not the control system (puts units on natural gas and sent oil sample out for analysis)
- Chilled water pump had impeller installed backwards!!
- Hot and cold deck AHU coils plugged with dirt (never cleaned!)
- Business Office "too cold"—(cooling for the entire floor is controlled by a "sensor" located in the hallway between the Accounting and Materials Management Departments!)

- Note: the engineer's office provides no privacy and the telephone isn't long-distance capable

And those were just the highlights! I think you'll agree, if ever there was a lady screaming for attention, this was she, but a mere hug and kiss wouldn't bring her around. She needed help...bad! Over the next several weeks, I continued putting out fires and logging in new flareups while I got better acquainted with the facility, its history and idiosyncrasies.

During that time, problems that came to light ranged from staffing & labor relations problems and system & structural failures to noncompliance with regulatory requirements and disgruntled administrators & department heads.

The temptation, of course, was to address all the problems as they became evident and provide immediate (though temporary) solutions to them. I resisted until, after approximately 6 months, I felt sufficiently knowledgeable of her quirks and passions to exact a more intimate probing of her nature and condition. What's that? The format? Oh, yeah ... here it is: feel free to modify it to fit your own particular situation.

Figure 2-1. Table of contents taken from a prior Facilities Evaluation report

```
I    EXTERIOR BUILDING/GROUNDS
     A  -  Existing Conditions
     B  -  Outdoor Utilities/Equipment
     C  -  Outbuildings/Satellites

II   PHYSICAL PLANT
     A  -  Interior Structures
     B  -  Equipment/System Overview
     C  -  Preventive Maintenance

III  PLANT OPERATIONS DEPARTMENT
     A  -  1989/90 Department
     B  -  Organization
     C  -  Operations

IV   REGULATORY/ENVIRONMENTAL ISSUES
     A  -  Asbestos Abatement
     B  -  PCB Contaminated Transformers
     C  -  Underground Oil Tanks
     D  -  Incinerator Air Emissions
     E  -  Other issues

V    SUMMARY
```

APPENDIX

1 Roof Survey
2 Underground Oil Tanks
3 Light Standard Survey
4 Power House Floor Survey
5 Wall Penetration Report
6 Bathroom Floor Tile Replacement
7 Door Survey
8 Generator Assessment Report
9 Incinerator Repair
10 Elevator Full Load Test Report
11 Dish Machine Repair Estimate
12 Electrical Distribution System Survey
13 Frozen Coil History
14 Preventive Maintenance Plan
15 Department Reorganization Inquiry
16 Request for Staffing Increase
17 Department Computerization
18 Work Order System Description
19 S.O.P. Outline
20 Reclassification of Personnel

REFERENCES AND REVIEW DOCUMENTS

Assuming you've expended adequate time to observe the goings on around you and acquired some modicum of understanding of your operations, "now," as the saying goes, "is as good a time as any" to give your facility the once over. But before you even consider venturing out of your office, first assemble and review:

- the facility's as-built drawings, project close-out documents, and fixed movable depreciation schedules
- the building's "statement of construction"
- all applicable building codes
- rules, regulations and licensing requirements particular to your organization
- a set of National Fire Prevention Association (NFPA) code books including the National Electrical Code (NEC) and the Life Safety

Code (NFPA 101)
- a copy of ANSI — "Specifications for Making Buildings & Facilities Accessible to and Usable by Physically Handicapped People"
- organizational hierarchy charts; a list of department heads' names and telephone extensions
- copies of all prior performed evaluations
- existing master lists
- outstanding recommendations from past inspections
- current interpretations of environmental legislation
- any other information you determine relevant to the assessment

ASSEMBLING A TEAM

Now that you've obliterated your desk top with reference materials, I'm sure the magnitude of your proposed undertaking is becoming painfully apparent to you. You're right, it'll take more than yourself and a few well chosen subordinates to pull it off. Significantly more! I can hear the questions rattling around in your head. How much of my time will I have to devote to the project? My peoples' time? When will they do their regularly scheduled work? Will the process cost me much in overtime premiums? What if I need to pull people for an emergency?

Well I have a couple of questions for you ... who says you have to tie your or your peoples' time up indefinitely to do a good job of assessing? Why not bring in some outsiders to help you? What do you mean, what do I mean? Let's face it; *any* time you devote to the evaluation is time taken away from normal operations. Robbing Peter to pay Paul only puts you deeper into the hole. My suggestion? Hire as much expertise as is economically feasible from the outside and limit in-house involvement to the orchestration of their efforts.

If properly focused, vendors who make their living in specialized trade areas are generally better qualified to make integrity judgements and can offer new insight into old problems. Too, due to their experience, they can home-in on difficulties long before in-house mechanics might make the same diagnosis, and their familiarity with similar systems and

Preparing for the Process

equipment predisposes them to recognizing potential trouble spots.

Though my last POE was comprehensive in scope and took 6 months to complete, I was able to accomplish it with a minimum of expense and in-house distraction. Many of the individual systems assessments, and survey summaries that appeared in the report's appendix were verbatim extracts of conclusions derived by outsiders who were hired or performed gratis inspections as a prelude to possible future employment. Some examples:

— thermographic and visual roof inspections/photographs, drawings & written report (no charge)

— visual and ultrasonic inspection of underground oil storage tanks/photographs, readings and written report ($125)

— outdoor building and grounds lighting survey/drawings & recommendations (no charge)

— floor loading study in the powerhouse by a consulting engineer/drawings, report and recommendations ($1,500)

— wall, floor & ceiling penetration study/daily reports and repairs (hired contractor on time and material basis)

— resistor bank electrical generators tests/tabulated results, opinions and written recommendations ($250/machine)

— assessment of incinerators/list of needed parts and written recommendations (brought in manufacturer's rep on an hourly rate)

— repair estimates on 3 dish machines/list of parts, cost breakdown and bid for the work (no charge)

— thermographic study of main and secondary electric distribution panels/photographs ($500)

— eddy current testing of chiller tube bundles/test results and written recommendation (no charge—part of service agreement)

— other studies/
  - elevator full load tests (part of contract)
  - interior finishes (sponsored by manufacturer)
  - ergonomic surveys (furniture outlets)
  - lighting/power factor studies (electrical manufacturers)
  - main switch gear inspections (utility company)
  - water treatment programs (chemical companies)
  - consumable materials inventory (product distributors)
  - department and tenant surveys (in-house)

The cardinal rule when availing yourself of these low- or no-cost studies is to establish (up front) with the providers that their performance in no way constitutes an agreement for future work. The utilities and your tenants aside, companies are well aware that they are in competition with other firms for your business, which is the reason for their having made such an attractive offer in the first place. Once you've decided what expertise you'll bring in, at what juncture of the evaluation, a staging area should be prepared to provide access to the reference materials, storage space for instrumentation, and suitable meeting quarters.

Surveys And Forms

Once an area or piece of equipment has been looked at, most of the in-house scrutiny of your physical plant can be performed by your own people who, paying a little more than usual attention to detail during their rounds, can funnel huge amounts of significant information back to you for your consideration.

The collected data can help you by alerting the mechanic to conditions requiring immediate attention, manifesting that similar problems exist in separate areas of the plant, providing base line information for making future comparisons, recognizing voids in standards and revealing the establishment of trends.

Another means of acquiring information from "upstairs," is through the use of questionnaires sent out to the building's tenants and/or departments requesting feedback on aesthetics, functionality and the provision of services. Regardless if you choose either or both, its important that you determine exactly what questions you want to ask and provide a suitable vehicle for documenting the answers. (See Figure 2-2, pages 22 & 23.)

TOOLING UP

You say the reference materials have been assembled in the ante room you've designated as the war room for the evaluation? Okay. You've constructed the data collection forms your people will be using and distributed all the tenant questionnaires? Great! And you've scheduled appearance dates for all the vendors you'll be utilizing? Marvelous...you're on the ball. Did you put a requested return date on the questionnaires? Uh-huh! Have you made provisions for assigning keys and supplying temporary ID badges to the outsiders? Will your people be escorting them around the building?

Perhaps you need to think this thing through just a bit longer. Besides those items we've already discussed you may want to consider:

- conducting a kick-off meeting
- formally notifying all affected departments
- compiling a schedule of shut downs and service interruptions
- pre-arranging for sample analysis with an independent laboratory
- using a video camera for documentation (and later training)
- calibrating existing diagnostic equipment
- utilizing a CADD system for updating drawings
- holding frequent de-briefings of all evaluation team members
- buying lots of batteries

Figure 2-2. Boiler operator guidelines indicate the need for a solid water treatment probram. *(Courtesy: Betz Laboratories)*

Boiler Type: Industrial watertube, high duty, primary fuel fired, drum type
Makeup Water Percentage: Up to 100% of feedwater
Conditions: Includes superheater, turbine drives, or process restriction on steam purity
Saturated Steam Purity Target[9]

Drum Operating Pressure[1] MPa (psig)	0-2.07 (0-300)	2.08-3.10 (301-450)	3.11-4.14 (451-600)	4.15-5.17 (601-750)	5.18-6.21 (751-900)	6.22-6.89 (901-1000)	6.90-10.34 (1001-1500)	10.35-13.79 (1501-2000)
Feedwater[7]								
Dissolved oxygen (mg/L O₂) measured before oxygen scavenger addition[8]	<0.04	<0.04	<0.007	<0.007	<0.007	<0.007	<0.007	<0.007
Total iron (mg/L Fe)	≤0.100	≤0.050	≤0.030	≤0.025	≤0.020	≤0.020	≤0.010	≤0.010
Total copper (mg/L Cu)	≤0.050	≤0.025	≤0.020	≤0.020	≤0.015	≤0.015	≤0.010	≤0.010
Total hardness (mg/L CaCO₃)	≤0.300	≤0.300	≤0.200	≤0.200	≤0.100	≤0.050	—Not Detectable—	
pH range @ 25°C	7.5-10.0	7.5-10.0	7.5-10.0	7.5-10.0	7.5-10.0	8.5-9.5	9.0-9.6	9.0-9.6
Chemicals for preboiler system protection						Use only volatile alkaline materials		
Nonvolatile TOC (mg/L C)[6]	<1	<1	<0.5	<0.5	<0.5	—As low as possible, <0.2—		
Oily matter (mg/L)	<1	<1	<0.5	<0.5	<0.5	—As low as possible, <0.2—		
Boiler Water								
Silica (mg/L SiO₂)	≤150	≤90	≤40	≤30	≤20	≤8	≤2	<1
Total alkalinity (mg/L CaCO₃)	<350[3]	<300[3]	<250[3]	<200[3]	<150[3]	<100[3]	—Not Specified[4]—	
Free hydroxide alkalinity (mg/L CaCO₃)[2]				Not Specified				
Specific conductance (μS/cm) (μmho/cm) @ 25°C without neutralization	<3500[5]	<3000[5]	<2500[5]	<2000[5]	<1500[5]	<1000[5]	≤150	≤100

NOTES

1. With local heat fluxes >473.2 kW/m² (>150,000 Btu/hr/ft²), use values for the next higher pressure range.

2. Minimum level of OH⁻ alkalinity in boilers below 6.21 MPa (900 psig) must be individually specified with regard to silica solubility and other components of internal treatment.

3. Maximum total alkalinity consistent with acceptable steam purity. If necessary, should override conductance as blowdown control parameter. If makeup is demineralized water at 4.14 MPa (600 psig) to 6.89 MPa (1000 psig), boiler water alkalinity and conductance should be that in table for 6.90 to 10.34 MPa (1001 to 1500 psig) range.

4. Not detectable in these cases refers to free sodium or potassium hydroxide alkalinity. Some small variable amount of total alkalinity will be present and measurable with the assumed congruent or coordinated phosphate-pH control or volatile treatment employed at these high pressure ranges.

5. Maximum values often not achievable without exceeding suggested maximum total alkalinity values, especially in boilers below 6.21 MPa (900 psig) with >20% makeup of water whose total alkalinity is >20% of TDS naturally or after pretreatment by lime-soda, or sodium cycle ion exchange softening. Actual permissible conductance values to achieve any desired steam purity must be established for each case by careful steam purity measurements. Relationship between conductance and steam purity is affected by too many variables to allow its reduction to a simple list of tabulated values.

6. Nonvolatile TOC is that organic carbon not intentionally added as part of the water treatment regime.

7. Boilers below 6.21 MPa (900 psig) with large furnaces, large steam release space and internal chelant, polymer, and/or antifoam treatment can sometimes tolerate higher levels of feedwater impurities than those in the table and still achieve adequate deposition control and steam purity. Removal of these impurities by external pretreatment is always a more positive solution. Alternatives must be evaluated as to practicality and economics in each individual case.

8. Values in table assume existence of a deaerator.

9. No values given because steam purity achievable depends upon many variables, including boiler water total alkalinity and specific conductance as well as design of boiler, steam drum internals, and operating conditions (Note 5). Since boilers in this category require a relatively high degree of steam purity, other operating parameters must be set as low as necessary to achieve this high purity for protection of the superheaters and turbines and/or to avoid process contamination.

Chapter 3
The Team Leader's Walk Through

I don't care how many people you send forth to poke, prod and ogle your facility; their efforts will be for naught if you can't understand or incorporate the information they provide you. And it doesn't matter how many times you've walked your halls or how many mechanical spaces you've stuck your head into, to check out a screech, click or thump. Prior to implementing a full-blown evaluation, it's imperative that you perform a thorough walk through inspection of your own.

Honing Your Homing Skills

Though it's often said, the truth be known, very few of us *really* know our facilities "like the back of our hand." Because buildings, especially those of considerable stature, are in a constant state of flux—tenants come and go, floors and walls are reconfigured, systems are rerouted, devices are replaced—even the normal wear and tear it endures and the aging process works against us.

As the project's team leader, you'll be responsible for assigning who will go where to do what. To be certain you're sending a qualified person to the right location to perform an appropriate task, you must first know the area and what can be found there. If questions arise, you'll need to be capable of visualizing the scenario presented in order to provide informed judgements.

Another of your duties will be to interpret the findings and determine the extent of need they place on your operations. These are formidable duties at least, which test your knowledge of the facility's layout, the location and foibles of its equipment, systems routes, zoning and operating history. Regardless of the scope of the evaluation, its relative success or failure hinges on you and what you know. So, ... what are you waiting for?

Figure 3-1. It's imperative you know in which direction you'll proceed.

There is no one way to go about conducting an orientation. The only rule is that you cover every square inch of the facility and that you feel confident you can recall what you've seen afterwards. You can start at the top and work your way down, vice-versa, visit all the mechanical spaces first or trace out individual systems ... it's up to you. What's important is that you cover the entire complex— personally.

During the course of your inspection, you may or may not want to drop into tenant occupancies for off-the-cuff discussions of their wants, needs and concerns; personally, I prefer to restrict these interviews until after they've filled out and returned my questionnaires. Unless you're saddled with too much of it (ha-ha), you'll want to limit the time you spend to viewing what you survey with a certain bent. Remember, at this stage, you're only preparing for the evaluation, not performing it. However you decide to traverse the place, note the following as you go:

- the size, general condition and contents of visited spaces
- the operating condition of and the areas served by equipment
- whether functional adjacencies seem logical
- lack of idiot lights and labeling
- blatant discrepancies
- obvious code violations

The Team Leader's Tool Pouch

No cheating! When I said every square inch, I wasn't exaggerating. Oh! You can confine your tour to the floor and leave the climbing to the grease monkeys but you've still got to visit those "far away places." How's that? Sure, there will be some tight squeezes and the lighting won't always be good, ... you may even suffer some minor abrasions and contusions along the way but then this isn't an ice-cream social you're attending, is it?

To help minimize the cuts, scrapes and singed arms, might I suggest a heavy pair of long-sleeved coveralls. To avoid bumps, bruises and stubbed toes—a hard hat and steel-lined shoes. And to keep your digits from becoming midgets, a quality pair of engineer's gloves. Some words to the wise from the voice of experience—remove all the jewelry from your person and paraphernalia from your pockets, take off your tie, tuck everything in and keep your eyes peeled—preferably behind a pair of safety glasses. Given you'll be suitably attired, here are some gizmos and gadgets you might carry with you to aide you in your venture:

- mini tape recorder
- battery operated light
- small inspection mirror
- tape measure
- camera
- heat gun
- small sample vials
- electrical tester
- screwdriver and pliers
- torpedo level
- wire brush and rags

COMPARING FACILITIES SERVICES

Every facility is laid out differently from every other one; even matching towers can't claim to be identical twins when held up to the light for close scrutiny—especially after they've been around for a while and acquired their own characters. But all facilities are alike in one respect—each is like a city in microcosm. You don't buy that? Think about it! Tell me one thing any municipality has that can't be had in a given facility. Better yet, here's a list comparing the services generally available there, to those that can be found in most of our operations:

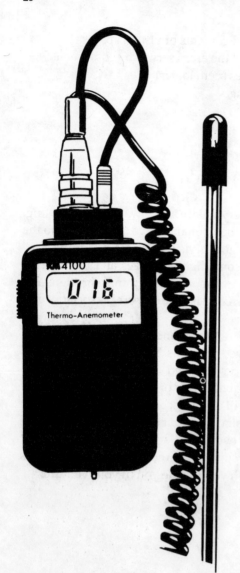

Figure 3-2. You may want to add this to your tool pouch for checking air velocities.

City Service
- electric power

- street lighting

Building Service
emergency generators, uninterruptible power supplies and transformers

outdoor standards and security lighting

- fuel supply — above- and below-ground fuel tanks
- water supply — hot and cold water supplies
- police protection — security guards, alarms and systems
- fire protection — fire pumps, sprinkler systems, alarms, and organized fire brigades
- telephone service — in-house telephone systems
- community broadcasting — teleconference rooms and two-way radio hook-ups
- trash removal — janitorial services, waste hauling contracts and/or on-site incineration
- snow removal — groundskeeping personnel and equipment
- water treatment — treatment of chilled water and boiler feed water/condensate loops
- repair crews — maintenance personnel

By no means do comparisons stop there; every storefront in town could conceivably be housed within the confines of a building where you can find restaurants in the form of cafeterias, gymnasiums scaled down to health clubs, outpatient clinics doubling as hospital emergency rooms and a plethora of other sundry sales and service entities. And, if you want to stretch the point, the building's owner or the CEO can be looked at as the Mayor, making the Board of Directors its Council ... and you?

A TYPICAL TOUR

Over the years, I've probably looked at my buildings from every conceivable vantage point. Say, what? Yes sir, even from underneath! That's why I submit that there really isn't any *right* way to tour them. Starting points are arbitrary. It's where you end up, not where you begin, that's important. As I established my roots in Uncle Sam's shipboard power plants, it seems I've developed a propensity for the boiler room and thus have always initiated my tours there. If you'd like to join me, I'm ready to commence my rounds from there now... Great—the experience will do us both good. Grab a pen and clipboard and we can be on our way.

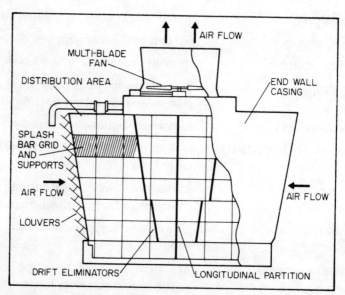

Figure 3-3. When touring the power house, don't forget to look on the roof.

THE POWERHOUSE

Aside from re-establishing my affinity for the Stationary Engineering professional (I are one), I've found that originating my rounds in the major mechanical spaces gives me a better feel for operations throughout the remainder of the plant. If things are going well here, there are likely fewer problems downstream. The powerhouse is the heart of the operation:

- on the lower level, sectioned off by concrete block walls are three high-pressure/dual fuel (natural gas/#6 oil) water-tube steam boilers, a cadre of pumps, two batch-type incinerators and the operating engineer's office which houses the energy management system computer.
- on the second level are four centrifugal chillers, main service shut-off valves for city water, chilled water and steam supplies and the boiler's deaerating feed water heater.
- on the upper level are located the main air handling units, pneumatic control system air-compressors and the remainder of the auxiliary support equipment

- finally, the roof holds the cooling towers, exhaust fans and incinerator stacks (the boilers have a separate free-standing brick stack).

Let's find out what, if anything, the operators have to report. "No news," as the saying goes, "is good news." Nonetheless, we shouldn't leave until we've determined:

- the overall condition of the power plant
- the capacity of the units
- how the equipment is loaded/unloaded
- if the equipment has redundant back-up or
- how it can be cross connected
- which equipment is normally up and on stand-by
- whether emergency contingency plans are in place

Figure 3-4. You may want to consider adding a light standard, here and there.

The Grounds

That out of the way, shall we patrol the grounds before entering the main building? Super ...

- to the left of the powerhouse is an attached brick structure housing two of the facility's four emergency generators; all of which are powered by #2 diesel fuel oil stored underground.

- to the right, near the boiler's stack, are the trash compactor, incinerator, access elevator and two #6 fuel-oil tanks buried underground.

- the water and natural gas mains both enter at the rear of the building, and the 4160-volt electrical substations are located at the opposite corner.

- all exterior lighting operates on timers or photo cells

How about taking a quick peek at the incinerator stacks and check them for smoke? None? Good! Some items to be cognizant of before leaving the grounds? Sure ...

- the frequency of generator operation and areas served
- the location of underground tanks and their contents
- the underground electrical feed route into the main building
- the lay out of the incinerator room and its history of down time
- the general condition of all paved areas and building facades
- the aesthetic appearance of the landscape
- the effectiveness of ways of travel and capacity of parking areas
- the location of fire hydrants and Siamese connections

Other Mechanicals

Needless to say, if it wasn't found in, on, or around the powerhouse or buried underground, it must be in the building proper. So, what are we missing? Tell me what they are and I'll tell you where to find them.

The Team Leader's Walk Through

- the main switchgear room (basement)
- electrical distribution closets
- hot water tanks (1st floor mechanical room)
- the other two generators (2nd floor interstitial space)
- fire pump (basement/east end)
- air handling units (many & varied per diagram)
- auxiliary mechanicals (54 rooms scattered/per room schedule)
- elevators (4 hydraulic & 18 cabled)

Did you remember the fuel tank which supplies the main building's generators? It's buried out back near the loading dock doors, as evidenced by the vent running up the side of the building there. Except for the elevator equipment that about covers electro-mechanicals in the main complex. We'll visit each roof, mechanical space and penthouse before touring the interior floor by floor. What to look for?

- location/appearance/condition of control panels
- areas served/main feeds/shut-offs
- booster pumps for water systems
- electrical load transfer switches (auto or manual)
- tamper switch on sprinkler main
- stepdown transformers
- telephone trunks

Now that we've acquainted ourselves with everything in the facility that whirls, whistles and whines, what do you say we take the elevator to the top floor and walk our way down to the basement? Agreed? Fine, press #34 please. Huh? Okay, after lunch it is.

FACILITY INTERIORS

You were right of course. I don't think we could have negotiated such a long trip as an interior walk through without first stoking our furnaces with a little fuel. Thanks for lunch! Ready? Let's go to the

Figure 3-5. Even older buildings can impress when well maintained. (*Courtesy of Weil-McLain*)

executive penthouse and work our way back, a floor at a time. This part of the inspection won't be as time-consuming as you might suspect. Except for movable partitioning, most of the floors are laid out fairly much the same and I prefer not to enter the tenants' private spaces unless invited.

We'll limit our sojourn to common spaces such as auditoriums, teleconference rooms and public ways of travel. Remember, at this stage we're mostly finding our way for future reference. While we're walking, commit the tenants' locations to memory and look for physical problems such as:

- obvious fire & electrical code violations
- stuck, broken or missing hardware
- vandalized finishes
- stained or missing floor/ceiling tiles
- cracked or broken window glass or trim
- loose railings & broken fixtures
- general housekeeping concerns

CHAPTER 4
DEALING WITH RISK

With every endeavor in life comes a measure of risk. Quarterbacks risk personal injury every time they're handed a football. The chances your car will be damaged increase each time you pull out of the drive. Bankers face financial insecurity with every loan they make. The fact is, even getting out of your bed in the morning doesn't guarantee you'll be around to slip back into it that night. Let's face it, if it exists, there's risk attached to it...be it legal, financial, health-related or what have you. And it follows that, if it can't be entirely avoided, the next best bet is to minimize one's exposure. This precept should be the guiding force behind every facility evaluation. Let me show you what I mean...

THE RISK PERSPECTIVE

Simply put, risk is a state of being wherein the possibility for loss or harm exists. In facilities, losses can be physical, such as the scrambling of a refrigeration compressor, inconvenience suffered by tenants as the result of service interruptions or reductions in the quality of goods produced, due to faulty equipment.

Harm, on the other hand, alludes to human compromises to health and losses of life and limb. Such outcomes can be consequences of the physical environment, utilization of improper or inadequate techniques and human intervention (or lack thereof). In any case, however risk may be manifested, all occurrences have financial implications which can ultimately affect your bottom line.

Every facet of an operation is at risk to some degree. Your job is to identify the perils, access their impairment potential, determine to what extent they can be tolerated and create a plan for their management. The areas of risk we'll cover in this text include general, electrical, life safety; security and regulatory compliance—each of which will be addressed separately in upcoming chapters. But before we delve into all that, you must first get a firm grasp on just what risk really is...

Figure 4-1. Considering the cloud formation above him, I'd say this guy has a lot to learn about risk.

THE CONCEPT OF RISK

Risks fall into two basic categories—speculative and pure. Speculative risks are those that organizations are subjected to which are difficult to predict or prevent, such as an act of God (lightning strikes, flooding) and acts of man (sabotage, terrorism) which in large part are dealt with through insurance coverages.

On the other hand, the outcomes of pure risks are somewhat more determinable or measurable and can be better planned for, as by providing back-up electrical power during outages, for instance. The sources of risk are four-fold, involving human, mechanical, environmental and procedural factors. "To err," as the saying goes, "is human." Human error is probably the single greatest contributor to risk exposure. But if man's ineptness *wins* the race, the unreliability of mechanical devices places a close second.

These two account for the bulk of the risks to be found in the physical plant. Environmental factors deal with both a building's physical (weather, temperature) and social surroundings (public mood, crime rates), each of which bears on or is borne upon by the first two factors.

Bringing up the rear are the facility anomalies resulting from the implementation and utilization of inappropriate policies, and/or procedures, rounds and routines. Procedural factors are the easiest and least expensive of the four to manipulate in effecting change.

DIAGNOSING EXPOSURE LEVELS

Whether you hire a so-called "risk expert," or decide to take on the task of determining how "at risk" your operations are yourself, it's important that you establish some means of keeping score as the evaluation proceeds. For each exposure, the scoring system you use must track each of the four risk factors we discussed earlier, determine their degree of involvement and assign values to them which, when tabulated, will reflect a pre-established level of risk. You can buy a canned program, suffer an expert or invent your own program. This is the scorecard I came up with to assess my own operations. Feel free to use it, if you like...

As the whole idea behind risk management is risk reduction and since my favorite pastime is golf, I thought it appropriate that achieving low scores should be reflective of success. My system is simple enough:

- Every area of exposure is scrutinized from the standpoint that, to some degree, each of the four sources of risk contributes to the overall risk level of the observed exposure.

- Values are assigned to each factor, totalled and compared to a risk level table.

- Risk levels are assessed and an action plan is drawn up to deal with the exposures.

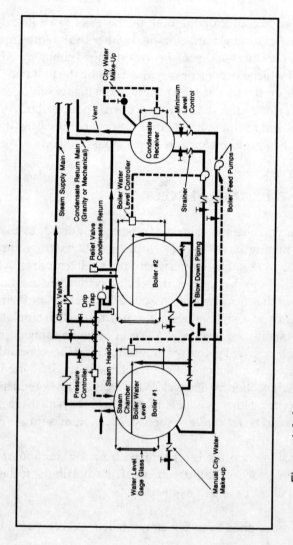

Figure 4-2. Low pressure boiler plant. (*Courtesy: Building Owners and Managers Institute*)

Dealng with Risk

Example #1

Determine to what degree a plant's steam boilers are at risk, using the following parameters:

Risk Valuations

0 — elements are correct, present & functioning
1 — elements are correct, present but non-functional
2 — elements are incorrect or not present

Risk Factors

Human	0	(Operators are licensed, competent and loyal employees who take great pride in their upkeep of the power house and equipment)
Mechanical	1	(though well maintained, the equipment is nearing the end of its normal life expectancy and suffers ever increasing incidents of repair)
Environmental	2	(equipment does not meet emissions requirements)
Procedural	0	(start up/shut-down and operating procedures are clearly written and conspicuously posted)
Total Score	3	

Risk Exposure

(Score)	(level)
0 - 2	low
3 - 5	meaningful
6 - 8	high

As can be seen by the final tally, even though the powerhouse is staffed with highly qualified personnel, working in accordance with stringent operating procedures, and the equipment is well cared for, a score of 3 places the boilers at a risk level (meaningful) which should prompt further consideration or action by the Facilities Manager.

Example #2

Using the aforementioned parameters, evaluate the effectiveness of the Company's Fire and Safety Committee:

Risk Evaluations

Source	Score	Notation
Human	2	(committee members show little interest in meeting agendas except when issues are regulation driven)
Mechanical	1	(furniture in meeting rooms is comprised of old school desks with fold-down arms)
Environmental	2	(meetings are held in a hot, noisy environment)
Procedural	0	(the committee members have drawn up and adopted an excellent charter for conducting safety proceedings)
Total Score	5	

A score of 5 also puts this exposure in the mea*ning*ful category but it's obvious that the committees capacity to produce is extremely compromised and in all likelihood will only deteriorate further unless corrective action is taken.

How the scoring is perceived and what changes, if any, are implemented as a result are functions of risk management. Decisions to act are based on numerous variables, including value, age, desirability, functional worth, and legal requirements among others. In the final analysis, you are the risk manager and it's important that you know how to ply your trade...

Managing the Threat

Risk Management is a process whereby a facility's potential for loss or harm is recognized, measured and acted upon. In some cases, the action taken is to take no action at all; in others, rigorous efforts often ensue. But not so fast. Before initiating the process, we must first consider...

Objectives

Regardless of their spans of control or the scopes of their operations, managers view their facilities with a common eye. Granted, some of us may deal with larger numbers or longer time frames than others but the risk-management process makes brothers of us all, whatever our title or station in the hierarchy. Our primary concern, of course, is to keep the operation operational; other objectives include:

- providing a safe environment
- averting catastrophic equipment failures
- avoiding unscheduled shut downs
- lowering operating costs
- complying with regulatory requirements
- minimizing systems malfunctions
- maintaining the building's appearance
- adhering to the corporate master plan
- "satisfying your customers"

Risk Distribution

Armed with a set of objectives, existing risks are more easily identified and measured, thus enabling decisions to:

1) Accept the risk
   (the action taken is to take no action, as the risk cannot be avoided, or is too difficult, time-consuming or expensive to correct)

2) Avoid the risk
   (by eliminating the source of the threat or danger)

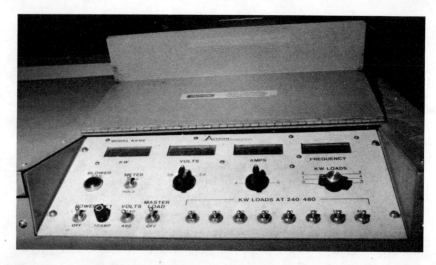

Figure 4-3. One way of averting catastrophic equipment failure is through a program of regular testing and calibration.

3) Reduce the risk
(by improving on the odds against its occurrence or minimizing its severity)

4) Redistributing the risk
(by reorganizing activities or changing procedures, such that risks are shared)

5) Transfer the risk
(through the use of insurances, service agreements and extended warrantees)

Controlling Strategies

What options are available to the risk manager to perform these risk manipulations? The alternatives to loss are as boundless as your imagination, your pocketbook and the time you can devote to implementing your resolves. Here are some "shuns" for your edificashun...

- separation   install barriers between hazards and those subject to harm

- evacuation — physically remove hazards from the site

- substitution — utilize less hazardous materials or techniques

- modification — improve the quality or characteristics of the end product/install redundant safety devices

- rehabilitation — make improvements on subsequent repair attempts

- prevention — anticipate and avoid dangerous conditions

- elimination — cease unsafe practices

- instruction — provide workers and end users with state-of-the-art training

- reduction — cut the amount of time or materials contributing to the risks/reduce vehicle speeds

- cooperation — provide an environment conducive to voluntary involvement by all persons affected

- immunization — design buildings and systems to withstand the onslaught of potential hazards

- clarification — post start-up/operating/shut-down procedures conspicuously by all equipment

TAKING CORRECTIVE ACTION

The preceding list of "shuns" is but a sampling of the strategies that can be utilized by Facilities Managers for correcting or controlling risks in their physical plants. Again, we are only limited by the extent of our common sense, training and available resources.

Using the two models, what conclusions might we draw from their exposure levels and how might the risks be distributed?

Example #1   Steam Boilers (exposure level - 3)

Even though a score of 3 is relatively low, two of the three points reflect an area over which the facility manager has no control; (i.e. the setting of emissions limits by jurisdictional authorities). This fact, in and of itself, pretty much dictates the action that will be taken. There are no choices here: reduce the risk; either correct the problem or be prepared to accept the consequences imposed by the authorities (shut down, fines and/or jail time).

Had the same score been arrived at, but two of the three points resulted from a lack of clearly written/posted operating procedures instead of a regulatory infraction, a whole nuther scenario presents itself. This being the case, it could be argued that the risk is minimized, if not negated by the competence of the operators on staff. Subsequently it may be decided to accept the risk or possibly transfer it by way of a Boiler & Machinery rider on the company's fire insurance policy. Whatever the decision, good engineering practice is to avoid the risk by eliminating the exposure. Conspicuously placed, written procedures are a good idea and an indisposable aid for supervisory types who operate equipment infrequently or the operator new on the job who hasn't yet become fully acquainted with the total operation. Besides, holding true to Murphy's Law, at the most critical moment the least amount of information can be retrieved from memory!

Example #2   Fire & Safety Committee (exposure level - 5)

While scrutinizing the committee, it was found that its then-present effectiveness was in question and its future success in doubt. Though not exhibiting a high-risk exposure level, it was painfully apparent that the committee's functionality was severely impaired and the implementation of corrective action was imperative.

Unlike the preceding example, the manager has a modicum of control over the environmental factors cited here. The exposure can be eliminated by moving the meeting place or improving conditions in the existing room and the mechanical exposure by replacing the desks with more suitable furniture.

Taking that action would drop the overall exposure level from 5 to 2, putting the meeting place into the low-risk range. Would your work be finished then? Not necessarily. Though a likelihood exists that the committee's members may gain a renewed interest in the issues they address as the direct result of an improved environment, there had to have been an inherent interest displayed in the first place.

How's that? Sure. What I'm driving at is how the committee was originally assembled. Did the people volunteer or were they assigned? Were they chosen based on their qualifications? Span of control in the plant? Concern about safety issues? Is the chairman the appropriate person for the job?

On the other hand, there are examples of groups saddled with meager surroundings who consistently improve their company's loss numbers, not so much as the result of their committee's responsibility to act as the corporate watchdog but from their individual commitments to safety consciousness in the workplace.

What do I want you to take away from this chapter? Just this: every aspect of every area needs to be considered for its potential for risk and the action necessary to manage it. The harder you work at lowering exposures—initially—the less likely you'll encounter catastrophe down the line.

CHAPTER 5
SAFETY MANAGEMENT

Safety. How do you define it? Security? Freedom from danger? A sailor's best friend? The penalty exacted for getting caught in your own end zone? I guess it depends on one's particular situation or concerns at the time they're concocting a definition. *Webster's* defines it as "...the quality or condition of being safe..." (I'm enlightened!). To my way of thinking, safety is the opposite of risk. As a general rule, the higher the degree of risk associated with an item, program or environment, the less "safe" it is.

Conversely, the safer these entities are considered to be, the lower the levels of risk they'll have associated with them. Safety issues encompass every facet of facilities management from equipment operation, building renovations and systems maintenance to fire protection, environmental compliance and pedestrian traffic flow. And I defy anyone to keep tabs on all of that without a viable plan to monitor and control it all.

PROGRAM CHARACTERISTICS

Before attempting to establish a safety program, you must first determine what your company's safety policy is, clarify it for every employee in the organization and endeavor to create an atmosphere of awareness throughout your facility. If the company has a good policy, leave it alone—nobody likes change. If the policy lacks teeth, accept it for the time being—it can be modified later. A non-existing or weak policy might be a blessing in disguise, allowing you the freedom to do things your way—assuming you can sell the people upstairs on your program. If your plan is to be successful, it should:

- conform to company policy
- have the backing and approval of the appropriate principles
- spell out who will be responsible for what
- possess input from all applicable departments
- be easily understandable
- encourage employee participation
- detail the components making up the corporate safety program

What components comprise a safety program? That depends on its scope, the level of safety you wish to attain and the resources budgeted for doing so. Generally speaking, a good safety program will include, at a minimum:

- the establishment of a safety department to implement it
- installation of a safety director to monitor it
- the appointment of a safety committee to suggest changes
- regularly scheduled safety meetings
- a system for incident reporting and accident investigation
- a technical library containing jurisdictional code books
- appropriate forms for filing and follow-up
- adequate time and space set aside for discussion of safety issues
- regular evaluation of the program
- conduction of hazard surveillance
- well thought out contingency plans for emergency operations
- safety orientation of new employees
- enforcement of a good housekeeping program
- comprehensive fire protection & security programs
- comprehensive electric safety & hazardous materials programs
- first-aid facilities and training
- guidelines for using protective clothing and safety equipment
- a handbook of company safety rules for each employee

Figure 5-1. Safety Program

Policy Number   _____

Effective Date _____

Revision Number _____

Page   _____

I.   Originator: Safety Director

II. Title: Safety Program

III. Scope: All Departments

IV. Purpose: To establish and maintain an optimal degree of safety throughout the facility

V.   Organization:

    The Safety Program is managed by the Director of Safety who coordinates and cooperates with the Safety Committee and the various department heads in developing and enforcing the policies of the Safety Program.

VI. Responsibilities of Director:

    1. Control and coordinate the Safety Program
    2. Responsible for loss prevention activities
    3. Develops and implements programs for safety including:

        a) Fire Safety
        b) Electrical Safety
        c) Safety Inspections
        d) Incident Reporting
        e) Disaster Preparedness
        f) Safety Committee

Figure 5-1. Safety Program (cont'd)

Policy Number _____

Effective Date _____

Revision Number _____

Page _____

4. Record keeping:

    a) Accident statistics
    b) Evaluation of the program with outside experts and inspecting agencies.

5. Safety training

VII. Responsibilities of Administration:

1. Legal responsibility for safety of visitors and employees
2. Administration efforts are directed toward the enhancement and support of the program.

    a. Safety Policy - to state Administrative policy and attitude toward safety.
    b. Safety Program Maintenance - support a structure for implementing safety policies.

VIII. Responsibilities of Department Heads:

1. Responsible for the absolute safety of all personnel in their areas to include the following:

    a. Safety Policy Development - Each department is to interpret, develop and expand the company's general safety policy into specific rules, policies and procedures relevant to their department.

Figure 5-1. Safety Program (cont'd)

```
Policy Number    _____

Effective Date   _____

Revision Number  _____

Page             _____
```

  b. Safety Policy Enforcement - To effectively carry out all
   policies by:
- Stressing to each employee the understanding that violations of established safety rules will not be tolerated.
- Continuous inspection for unsafe practices and conditions.
- Prompt corrective action to eliminate safety hazards
- Supporting fully all safety activities and policies

  c. Safety Training - To train each employee as to what hazards are on the jobs and how to avoid them. Complete safety instructions are to be given to all employees prior to assignment of duties.

IX. Responsibilities of Employees:

  1. Safety Compliance - Employees are expected to follow safe procedures and to take an active part in the work of protecting themselves, visitors, their fellow workers and company property. Everyone must know their exact duties in case of fire or other emergency. The objective is to make safety a part of the job accomplished by following:

    a Company and department safety rules
    b) Taking no unnecessary changes
    c) Using all safeguards and safety equipment provided
    d) Cooperating in every respect with all safety programs

Figure 5-1. Safety Program (conclusion)

Policy Number   _____

Effective Date   _____

Revision Number _____

Page   _____

    2. Safety Reporting - Employees are expected to detect and report to their supervisors all hazardous conditions, practices, and behavior in their work areas and to make suggestions for their correction. In cases of injury, employees are required to report promptly an injury, regardless of its seriousness, to the department head or supervisor. An incident report should also be filed, for failure to report the injury could result in forfeiture of Workmen's Compensation benefits.

X.   Safety Education and Training

    1. The Company will provide continuous programs of education on safety using films, posters, signs and meetings. Among others, subjects may include:

       a) fire prevention and fire safety
       b) sanitation, environmental safety
       c) lifting methods and proper handling
       d) electrical safety
       e) visitor safety
       f) disaster plans
       g) employee orientation

    2. Posters and signs will be utilized as reminders to personnel of their role in assuring a safe environment.

_____
Safety Director   Date

_____
Administrative Approval   Date

Safety Management

THE SAFETY COMMITTEE

Whether your organization is large enough to support a full-fledged, stand-alone department or chooses to address the safety issue by another means, every facility can benefit from the presence of a safety committee comprised of members selected from all areas of the operation. And membership needn't be restricted to the upper echelons of the management ranks ... after all, a high level in the hierarchy doesn't necessarily reflect a high understanding of safety. Besides, I've always held that it's the sergeants who are most knowledgeable of goings on in the field, the officers making their best decisions based on the input they receive from them.

However staffed, the committee must prevail upon itself to adopt a charter which defines its function and spells out the duties of its membership. Though there's no such thing as a standard safety charter, I've inserted a model in the appendix for you to study.

If you're from a health-care institution you'll doubly benefit as it was drawn up for a medical facility. If you're not, no matter; it's the form rather than the substance that's of significance here. Perusing the document you'll find:

- a definition of purpose
- the philosophy of the committee
- a statement of the committees authority
- structure of the membership
- the committee's objectives
- reference to adopted standards
- general responsibilities of the members
- specific duties of the chairperson
- specific duties of the secretary

POLICIES AND PROCEDURES

If a safety program is to pay dividends after its implementation, a significant, concerted effort must first be expended to properly establish it. As no system can survive long without rules, organizational safety

Figure 5-2. Writing a policy to cover equipment installation can be as time-consuming as the installations themselves.

policies and the procedures for carrying them out, must be formulated, implemented and enforced. Referring back a chapter, there are four sources of risk—human factors, mechanical items, environmental concerns and procedural problems. These are the areas which must be addressed if the program is to succeed. As you'll see, one area may well overlap into another. Here's a generic list of policy/procedural topics applicable to each heading:

Human Factors

- employee safety
- evacuation routes
- personal protective equipment
- safety training
- infection control
- communications
- right-to-know laws
- vandalism
- insubordination

Mechanical Items

- electrical safety
- fire system inspections
- equipment installation/removal
- system interruptions
- fire extinguisher types/locations
- lighting intensities
- alarm systems
- care & storage of tools
- calibration of instruments
- water treatment parameters

Environmental Concerns

- smoking
- asbestos management
- building and grounds
- inclement weather
- hazard reporting
- industrial hygiene
- storage of flammables
- hazardous waste disposal
- air-handling guidelines
- exhausting of harmful fumes

Procedural Problems

- safety training
- emergency preparedness
- fire plan
- accident/illness reporting
- materials handling
- code compliance
- recalls/consumer alerts
- duties of fire brigade
- breaches in security
- start-up/shut-down procedures

Figure 5-3. Company Safety Policy

Policy Number _____

Effective Date _____

Revision Number _____

Page _____

I. Originator: Safety Director

II. Title: Company Safety Policy

III. Scope: All Departments

IV. Purpose: To instill the Company's desire to provide a safe environment for all who pass through its portholes and establish a base regard for the issue of safety for all its departments to build on.

V. General:

1. Know your own departmental safety rules.
2. Eliminate horseplay and practical jokes.
3. Understand the safety requirements of your job.
4. Think about what you are doing and how you are doing it.
5. Report any unsafe conditions you see to your supervisor.
6. Suggest safe ways to perform your work.
7. If you are ill, do not come to work but call your supervisor to report off, giving as much notice as possible.

VI. Walking-working Surfaces

1. Practice good housekeeping.
2. Keep aisles and stairways free from obstruction.
3. Clean spilled liquids off the floor
4. Position electrical cords out of the way of traffic.
5. Walk, don't run.
6. Keep to the right and slow down at intersections.

Figure 5-3. Company Safety Policy (cont'd)

```
Policy Number    _____
Effective Date   _____
Revision Number  _____
Page             _____
```

      7. Open doors carefully
      8. If you have to climb, use a ladder. Don't use a box or chair.
      9. Never use a broken ladder
    10. Report defective equipment to your supervisor.

VII.   Means of Egress (Exitways)

      1. A safe, unobstructed way of exit travel from any point in the building to the outdoor ground level shall be maintained.
      2. Work will be carried on in such a way that obstruction of corridors, doors, stairs and other parts of exitways will be minimized. Carts and other materials will be placed near the wall on the same side of the corridor.
      3. If an exit must be blocked temporarily, place substantial signs or a guard at intersections to show alternate routes.

VIII.   Occupational Environment

      1. Clean up any spills, debris or residue from construction projects. If unable to do so immediately, block off the area to prevent injury.
      2. When entering special areas such as isolation rooms, observe the precautions for dress to protect others and yourself. Any tools or equipment used in such areas should be cleaned and treated as required to prevent cross contamination.
      3. Keep noise to a minimum.

IX.   Personal Protective Equipment

      1. Wear safe clothing. Torn sleeves, torn and oversized uniforms are dangerous. Wear shoes that are well built and in good repair.
      2. Wear the proper clothing for the job.

Figure 5-3. Company Safety Policy (cont'd)

Policy Number   _____

Effective Date   _____

Revision Number   _____

Page   _____

3. Wear safety goggles or glasses whenever there is a possibility of receiving eye injuries from flying particles.
4. Wear rubber gloves, aprons and face masks when working with cleaning solutions, acids, etc.
5. Wear respiratory equipment when working in operations which could produce accumulations of dust or dirt injurious to the respiratory tract or in which heat could release metal fumes or toxic vapors.

X. General Environment

1. Observe warning signs.
2. Place warning signs when work is being performed where it creates a hazard to other parties.
3. Barricade or block off the area if openings exist in floor or walls.
4. Use tags to warn of temporary hazards. Observe them.
5. Dispose of trash and waste properly.
6. Use established washing and toilet facilities.
7. Eat only in those areas so designated.

XI. Medical and First Aid

1. Report all injuries no matter how slight; secure immediate first aid and make certain an incident report is prepared.
2. All accidents, no matter how minor, are to be reported and an incident report completed.

XII. Fire Protection

1. Know the fire and internal disaster plan
2. Know the location of the nearest evacuation routes and of the nearest fire extinguishing equipment.
3. Follow smoking policies and adhere to all "No Smoking" signs.
4. Do not allow rags, and other combustibles to accumulate.

Safety Management

Figure 5-3. Company Safety Policy (conclusion)

Policy Number _____

Effective Date _____

Revision Number _____

Page _____

XIII. Material Handling

1. Never lift any object too heavy or bulky to lift comfortably or safely alone. Get help.
2. Use lifting equipment when working with heavy loads or weights and at heights. When working on ladders where lifting is involved, have a helper.
3. Use proper body positioning when lifting.
4. Do not stand under loads being handled by hoists.

XIV. Machinery

1. Use and maintain proper guarding for all machinery.
2. Place "Out of Order" tags on malfunctioning equipment.

XV. Portable Tools and Equipment

1. Keep all hand tools in safe condition.
2. Keep cutting tools sharp.
2. Use only non-sparking tools around flammable / explosive vapors.
3. Check extension cords for power tools carefully before use.

_____

Safety Director   Date

_____

Safety Committee   Date
Chairperson

_____

Administrative Review   Date

Department Safety Rules

Ideally, an organization's safety policies can be applied on a company-wide basis but those of us coming from the real world know that blanket rules aren't capable of covering every individual situation that might present itself. Subsequently, corporate policies should be general in form and text, allowing for specific addressing of safety issues at the department level.

The samples on pages 63-66 are typical of the safety policy commonly found in the average building maintenance department.

When new employees first join an organization they are unfamiliar with the layout of the physical plant, unaware of company policy and often unschooled in the area of safety. In a nutshell, accidents waiting to happen. Rather than allow them to run amok through the facility disrupting the safety savvy you've worked so hard to attain, they should be subjected to an intensive and informative safety orientation session.

Its main presenter should preferably be the company's director of safety or a principal of the Safety Department, supplemented by expert speakers from various departments to address the specifics of their operations. Such sessions typically cover:

- a review of the corporate hierarchy
- functions of the Safety Department/Director/Committee
- an overview of the company safety policy
- the employees' role in plant safety
- issues pertinent to the employees' area of work
- an explanation of the fire safety program
- the actions to be taken by employees during emergencies
- handling and storage of hazardous materials
- accident and incident reporting
- the employees' "right-to-know"

Figure 5-4. Department Safety Policy

        Policy Number   _____

        Effective Date   _____

        Revision Number   _____

        Page   _____

I. Originator: Director of Maintenance

II. Title: Department Safety Policy

III. Scope: All Maintenance Personnel

IV. Purpose: To establish safe working practices within the Department

V. Areas of Responsibility:

The Maintenance Department is responsible for the safe operation of all equipment in its charge regardless of its location within or outside of the building proper and for providing a safe environment for all personnel frequenting the buildings and grounds, be they employees or visitors.

VI. Maintenance, Operation, Testing and Inspection:

All activities involving the manipulation of any control, device, equipment or system component will be performed in accordance with pre-determined department schedules and procedures.

VII. Safety Training:
All personnel are required to attend the Company's safety orientation session before beginning work in the department, read and sign off on the department safety manual and familiarize themselves with the proper techniques for lifting and carrying materials and the safe use of hand and power tools.

Figure 5-4. Department Safety Policy (cont'd)

Policy Number   _____

Effective Date   _____

Revision Number   _____

Page   _____

VIII. Operation of Power Tools:

1. Power tools shall be inspected before use and only operated by authorized personnel, trained in their use.
2. When using power tools in wet areas, or when contacting metal surfaces, they must be electrically grounded
3. Any tool capable of creating a spark shall not be used in explosive or dusty atmosphere.
4. Electrical extension cords must be properly sized to carry the current of the tool it is supplying, be color coded and maintained such that it will not pose a hazard to personnel travelling through or working in the area.
5. Appropriate personal protective equipment must be worn whether the power tools utilized are portable or stationary.
6. Power tools must be maintained per manufacturers' instructions and properly stored when not is use.
7. Power tools shall never be energized and left unattended.

IX. Hand Tools:

Hand tools may only be used for their intended purposes and shall be discarded when worn or broken.

X. Machine Guards:

Equipment shall not be operated with the guards loose, broken or removed.

XI. Flame Cutting, Welding ad Soldering

1. Acetylene gas tanks will be capped at all times when not in use. They will be stored upright and secured with a chain or other

Figure 5-4. Department Safety Policy (cont'd)

Policy Number   _____

Effective Date   _____

Revision Number   _____

Page   _____

form of holding device. They will be kept from heat and flame, in a space designed for flammable gas storage.
2. Flame-cutting and welding will be accomplished in maintenance areas only, unless otherwise approved.
3. Personnel will wear proper protective equipment, including gloves hoods, goggles, aprons and ankle-high shoes, with trousers secured to prevent molten material from falling into the shoes.
4. Adequate screening and warning devices will be set up to prevent eye injuries to workers nearby.
5. Personnel should wear gloves and either goggles or a face shield while welding and soldering; keep their sleeves rolled down, shirt collars buttoned, and trouser legs over the shoe tops.
6. All containers containing explosive vapors must be evacuated prior to begining work

XII. Electrical Safety:

1. Prior to working on electrical circuits they must be de-energized, locked out and tagged. The circuits may only re-energized and the tags removed by the person initiating the repair.
2. All new installation work must be performed in accordance with the National Electrical Code.
3. Electrical repairs may only be performed by qualified personnel as assigned.

XIII. Paints and Solvents:

1. No volatile substances may be used in any area unless it is well ventilated
2. All combustible substances must be kept in UL approved safety containers
3. Smoking is not allowerd anywhere in the building except in areas designated by the Safety Director
4. Appropriate respiratory equipment must be worn when working in mists or high dust areas
5. No spraying may be done in areas where sparking or open flames exist.

Figure 5-4. Department Safety Policy (conclusion)

Policy Number _____

Effective Date _____

Revision Number _____

Page   _____

XIV. Working from a Height:

1. Ladders and scaffolding shall be kept in good repair, inspected and properly secured prior to use.
2. All walking surfaces shall be rendered and maintained in a "non-skid" condition
3. Proper techniques must always be used when positioning and using ladders in high traffic areas
4. Metal ladders are never to be used for performing electrical repairs

_____

Originator   Date

_____

Safety Director   Date
Approval

_____

Administrative   Date
Review

The Supervisors' Role

Front line supervisors are best positioned organizationally to run herd over safety issues. As the eyes and ears of management, they see into and hear about every aspect of the operation, first hand. Their constant contact with the rank and file sensitizes them to the needs of their subordinates and their never-ending facility tours make them cognizant of buildings and grounds discrepancies.

And let's face it: safety doesn't just happen ... someone, or something has to make it happen. No matter how well equipped or trained employees may be or what level of responsibility they've been assigned, it can't be assumed that safety will always be at the forefront of their thought processes. Consequently, someone has to oversee that facet of their employment.

Aside from their other myriad duties, the supervisor's role in safety is two-fold—first, as the company watchdog, keeping constant vigil over the company's physical assets, and second, as a champion of safety awareness for all persons involved in or affected by its operations.

These ancillary responsibilities are best accomplished with formal checklists used to uncover problems and aid in the training process. I've included a few in the Appendix for your perusal.

Chapter 6
Fire Protection

Fire, like small dogs and long-standing relationships, is one of life's realities we tend to take for granted, until it turns mean and bites us. As with everything else in the world, it has a good side and a bad one. I'm sure you share my sentiments when I say that "... the only good fire is a friendly fire." Huh? No kidding ... there really is such a thing as a friendly fire: the fire in your hearth at home is a good example ... at work it's the one supplying a source of thermal energy for comfort heating, food preparation or the manufacturing process. As facilities managers, our job is to make certain the "friendly" fires remain friendly, thereby protecting our buildings' occupants from suffering the consequences imposed by the other variety.

Fire Codes/Regulations

As a general rule, fire regulations throughout our country are generally not written by the jurisdictional authorities who enforce them. Rather, state and local fire codes are established through the adoption (in part or in whole) of the National Fire Codes. Though it has no regulatory authority itself, the National Fire Protection Association publishes a multi-volume set of codes annually which recommends up-to-date, pertinent fire-prevention standards and practices for use by all organizations. Without quoting chapter and verse, you might want to consider these areas during the course of your facility evaluation:

- the design construction and equipping of buildings to comply with all applicable codes
- a comprehensive statement of construction and fire – protection maintained on file for each building and frequently updated by a professional architect or engineer
- electrically supervised/manually activated fire warning systems which alarm locally and automatically notify organized fire departments
- automatic fire suppression systems
- proper sizes, types and locations of portable extinguishers and other fire fighting equipment
- automatic sprinkling systems/smoke and heat detectors
- organized in-house fire brigades
- policies and programs which promote fire-prevention and train personnel in the proper handling, storage and disposal of flammable solids, liquids and gasses
- provisions for back-up electrical power for egress illumination during emergencies
- installation and testing of fire safety systems and components

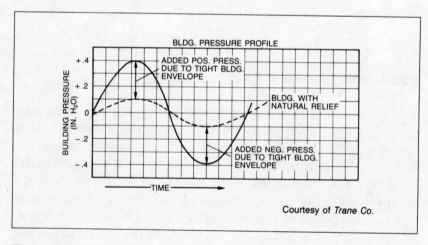

Figure 6-1. Fire suppression efforts are borne upon by which way the wind blows.

The Company Fire Program

Whether the construction and monitoring of the master fire plan falls squarely on your shoulders, has been assigned to the Safety Committee as one of their adjunct responsibilities or if a separate department was built around the need for having one, I guarantee you'll be involved with it from the day it's drawn up ... 'till the building burns down.

Figure 6-2. Fire Safety Program

```
                              Policy Number    _____

                              Effective Date   _____

                              Revision Number  _____

                              Page             _____
```

I.   ORIGINATOR: Director of Safety

II. TITLE: Fire Safety Program

III. SCOPE   All Departments

IV. PURPOSE   To provide a fire-safe environment for all occupants of and visitors to the company's properties, by maximizing fire safety awareness through training and monitoring of its personnel and equipment and the implementation of comprehensive fire prevention methods and measures.

V.   Organization:

The fire safety program is managed by the Director of Safety who coordinates with the Safety Committee and the various department heads in developing and enforcing policies.

VI. Responsibilities of Director:

   1. Control and monitor the program

   2. Responsible for loss prevention activities

   3. Determines and enforces company fire safety policy

Figure 6-2. Fire Safety Program (cont'd)

Policy Number _____

Effective Date _____

Revision Number _____

Page _____

4. Develops and implements training programs

5. Maintains records of fire incidents

6. Evaluates physical plant fire-safety status with outside experts and inspection agencies.

VII. Responsibilities of Administration:

1. Legal responsibility for the safety of visitors and employees

2. Support of the program and the Director of Safety in carrying it out

VIII. Responsibilities of Department Heads:

1. Each department is responsible for interpreting the intent of the company's fire safety policies and to expand on them in developing specific department rules, policies and procedures relevant to their operation.

2. Each department is responsible for continuously monitoring its operations to detect and correct unsafe practices, eliminating fire hazards and stressing the importance of fire-safety compliance to its employees.

IX. Responsibilities of Employees:

1. Attendance at all fire training programs

2. Observance of department and company fire-safety rules of conduct

3. Proper use of the safeguards and safety equipment they are provided.

Figure 6-2. Fire Safety Program (cont'd)

> Policy Number _____
>
> Effective Date _____
>
> Revision Number _____
>
> Page _____

X. Safety Education & Training

   1. The Safety Department will continuously develop and conduct fire-safety educational presentations for company employees.

   2. The Director of Safety will conduct and critique fire drills involving each of the company's departments, once a year per established policy.

   3. Posters, signs and bulletin board notices will be utilized as reminders to personnel of their roles in assuring a fire-safe environment.

XI. Policies and Procedures

   The Director of Safety will write, implement and enforce fire-safety related policies and procedures covering areas such as:

   * conductance of drills

   * fire brigade organization

   * training methods

   * fire detection and suppression

   * company smoking policy

   * acceptable construction

   * electrical fire protection

   * fire-safe procedures

   * bomb scares

   * evacuation procedures

   * contingency plans

Figure 6-2. Fire Safety Program (conclusion)

Policy Number	_____
Effective Date	_____
Revision Number	_____
Page	_____

XII. Fire Brigade

A fire brigade comprised of selected members of individual departments will exist under the direction of the Safety Director who will provide for their training and organization.

XIII. Fire Drills

Fire drills will be conducted in accordance with Section X of this document.

_____
Safety Director   Date

_____
Safety Committee   Date
Chairperson

_____
Administrative   Date
Review

Life Safety Code History

For over 50 years, the N.F.P.A. (National Fire Protection Association) has published the Life Safety Code (NFPA 101). Known formerly as the Building Exits Code prepared by the NFPA Committee on Safety to Life, it addresses the warning and egress requirements having direct influence on life safety issues in new and existing structures.

Appointed in 1913, for its first few years the Committee studied notable fires involving loss of life, analyzing their causes. From their findings, they developed standards for the construction and arrangements of stairways and exit facilities, forming the basis of the modern code.

In 1921, the Committee pooled the data they had accumulated to that date and prepared the Building Exits Code, a comprehensive guide to exits and related features of life safety from fire in all classes of occupancy. Every couple of years, the code was refined and a new edition was put out.

In 1966, a complete revision of the 1963 edition was accompanied by a change in the code title from Building Exits Code to "Code for Safety to Life from Fire in Buildings and Structures," and the text was translated into code language putting all explanatory notes into an appendix for compatibility with and use as a building code supplement.

In 1967, the Code was placed on a 3-year revision cycle. In 1977, the Committee on Safety to Life was reorganized as a Technical Committee comprised of an Executive Committee and 11 standing subcommittees responsible for various chapters and sections of the Code.

Today, the NFPA is comprised of more than 175 technical committees involved in standards-making. The Committee on Safety to Life consists of experts in the design and construction of buildings and structures, the manufacture and testing of building components and accessories and the enforcement of Life Safety related regulations.

The primary impetus of the Code deals with the safety of life as opposed to the protection of property, though as a matter of course the first obviously benefits the second. The lives saved in applying the Code is a number that's hard to determine, but it's reasonable to assume from studying its past application, or failure thereof (in cases of loss), that the Code has had and is having a profound effect on the preservation of life.

Structural Containment

Basic to the concept of protecting life from the consequences of an unfriendly fire is the avoidance of the fire and its products of combustion. And the best way of achieving that end in a structure is through physical separation by room, corridor, smoke compartment, floor assembly or building structures, each of which provides a distinguishable level of fire protection. Known as the "unit concept," fire and smoke is contained via compartmentation which employs consecutive levels or *units* to impede its progress through a building or structure.

The first level of defense is the room. Ideally, each room should be a totally sealed enclosure and be capable of protecting its occupants from the spread of fire and smoke for a predetermined length of time depending on its function and the materials and substances it contains, as outlined in the building's Statement of Construction. Openings such as for doors, windows, etc., must, of course, be compensated for. A room's fire and smoke integrity depends on its component wall, floor and ceiling structures, interior finishes, control of its openings and engineered ventilation.

The next level of defense is the smoke compartment which can be comprised of a number of rooms and passageways (as opposed to fire corridors). Smoke compartments stop smoke penetration from other areas and are used for evacuating persons from smoke-filled areas on the same floor or level of a building. To afford this protection, they must be continuous from floor slab to floor (or roof) slab above and from outside wall to outside wall, continuous through all concealed spaces. Doors in smoke partitions must be self-closing and duct work must be protected by smoke dampers, both of which react concomitantly upon activation of the building's fire warning system.

Floor assemblies are a structure's third line of defense, functioning as both fire and smoke barriers from the stories above and below it. The penetrations found in floor assemblies of multi-story structures are numerous and require special consideration during their design or retrofitting. They include stairwells, chutes, elevator shafts, utility chases and the occasional hole knocked in the wall for ... God knows what! Such designed openings must be protected by vertical partitioning or otherwise compensated for. The spaces surrounding pipe and duct penetrations should be completely sealed with fire-resistant materials, as would the exploratory holes your mechanics made. (You might want to ascertain if the holes they made were in bearing walls, and install a lintel or two while you're at it!)

Level four involves the building proper. It must be capable of remaining intact under conditions of fire and to contain the fire within itself. The time period for which the building must withstand the fire is determined by its height, and the Life Safety Code requires fire walls between different occupancies of buildings of different construction types. But these concerns lie more with the architect than with the Facility Manager. Your job is to make certain that the architect followed the proper code edition and the contractor met the specifications at the time the edifice was constructed.

The last level or unit of defense against the physical passage of fire and smoke through a structure is the exit. Your job here is to determine if the right number of exits exist, that they're the proper distance from all interior points, that they can be easily identified, if their doors swing in the right direction, and that the way of travel to and through them remains unimpeded.

LIFE SAFETY SYSTEMS

There are three major overlapping categories of life safety systems: construction, which we've already discussed under the heading of structural containment; building systems, such as emergency power circuitry; and fire-related systems used in fire detection and suppression.

The second category, autonomous electrical systems, affect fire safety through illumination of the means of egress, directional and exit signage, security lighting, activation of fire, and sprinkler alarms and electrical energizing of fire pumps, elevators, engineered smoke exhaust motors and critical life safety branches of electrical circuits.

The third category, fire related systems, can be subdivided into the areas of detection & warning, signalling and extinguishment. Early detection & warning can be achieved manually, such as through the use of pull boxes or automatically, as with the use of smoke and heat detectors hard-wired into a monitored fire warning and suppression system. Detection systems provide early warning of fires and often serve to alert in-house personnel responsible for responding to fire emergencies, to the location of the potential conflagration, allowing fire-brigades to contain the fire until the local fire department arrives.

Signalling of the municipal fire department can also be performed manually or automatically, but certainly it's preferential to have them notified automatically at the moment the system initially detects a fire, before the fire has a chance to intensify to a level of heavy smoke and high heat generation. Two common methods of notification are through a hard-wire tie-in of the building's fire-warning system into the fire department's in-house panel or via pre-recorded telephone messages transmitted to the department upon activation of the system.

Extinguishing systems include fire department connections, standpipes, sprinkler systems, fire extinguishers and automatic fire suppression systems installed in areas such as computer rooms and over cooking surfaces. To a degree, each of the life safety systems is mutually dependent and none of them can completely offset the need for the others.

Figure 6-3. Are your oxygen and combustible storage tanks safety located?

STORAGE OF FLAMMABLES

Whether acquired by the Facilities Manager or some other department head, as is the case with most bought materials, flammable items are normally bulk purchased to take advantage of available cost reductions or to maintain supply inventories. Unfortunately, this is often the last time their existence in quantity is considered. From the time they arrive at the loading dock until their final consumption, flammable substances and materials require extraordinary attention in their handling, transport, storage and use.

Generally speaking, unless some forethought is given to their layout and housekeeping, storage areas are largely unattended, often cluttered, heavily fire-loaded accidents waiting to happen. And, if you know Murphy as well as I do, you know the wait won't be long. Some points of import to consider in keeping Murphy at bay:

- the storage facility should be of proper size and engineered and constructed for the items and materials stored there

- the items stored there should be labeled to allow easy identification of their contents

- the height and arrangement of items should not interfere with the testing or functioning of fire-detection or fire-extinguishing system components

- solvents or other volatile liquids should not be stored in open containers

- aisles should be maintained free and clear of clutter, and shelves should not be overloaded

- trash bins in the area should be emptied regularly

- smoking should be prohibited as well as the use of anything that can generate a spark

- use containers specifically designed for handling volatile liquids, such as safety disposal cans, covered bench cans, or those having spring-loaded closures

- keep flammable liquids separate from other combustibles in a room especially designed for their storage or in otherwise approved cabinets

- provide protective clothing and equipment for handling solvents and install an eye wash station in close proximity to work areas

- use racks, pallets, skids and other equipment designed for quick, safe and easy materials handling

- maintain an inventory and discard all materials and substances that have exceeded their shelf life

- limit the transportation and use of solvents to a 1-day, properly containerized supply

- make certain storage areas are well lighted and posted with the proper signage

- visit the area frequently and make certain safety policies and procedures for the area are strictly observed.

Chapter 7
Electrical Safety

If fire, as I pointed out in the last chapter, is like a small dog waiting to bite you, electricity is like a snake in hiding, just aching to strike out at you, fangs first. And many are the individuals who can validate that observation. Certified electricians, who daily work with the serpentine aluminum and copper strands, admit to being struck from time to time. And even some of the most expert from their ranks have lost their lives as a result of a fleeting miscalculation. The bottom line is: never turn your back to an energized circuit and keep your eyes focused on...

Elements of Electrical Shock

Electrocution isn't just a ceremonial act exacted on the criminal element among us; it also occurs unceremoniously, involving us good guys, every day. As our society becomes more and more "electrified" chances are more people will be electrified proportionally, due to the increase in the number of electrical appliances surrounding them, the lack of adequate or proper maintenance or physical contact with energized circuits. And it doesn't take a lot of juice to cause you harm, as it's the amperage, not the voltage, that does you in. Current, delivered in small amounts at 60 cycles, can have these adverse effects on the human body.

Amperes	Effect on The Body
0.001	1 Milliampere. Threshold of perception: tingling sensation.
0.016	16 Milliamperes. "Can't-let-go current." Level of current where subject is unable to release grip on the electrical conductor.
0.050	50 Milliamperes. Pain, possible fainting, exhaustion, mechanical injury. Heart and respiratory function continue.
0.100	100 Milliamperes to 3 Amperes. Probable heart failure.

The natural resistance of the human skin is compromised as follows:

Resistance Value	Square Inches of Exposed Skin
400,000 OHMS	(0.5) dry calloused finger tip touching a dry conductor
15,000 OHMS	(0.5) dry calloused finger tip touching wet conductor
8,000 OHMS	(3) dry hand
1,600 OHMS	(15) dry hand
1,000 OHMS	(15) wet hand
100 OHMS	(15) hand blistered by 3-second flow of 50 milliamperes through hand

(I'll let you do the math!) —K.P.

Electrical Safety

Figure 7-1. Testing circuitry can be as complicated as the actual circuit wiring.

Program Considerations

Although it's perfectly permissible to incorporate electrical safety into an organization's all-inclusive corporate safety plan, I advise against the practice... if for no other reason than because it would eventually get lost in a financial resources shuffle. Unless yours is a warehousing operation, your facilities are extremely small, or your company's operations are completely lacking in sophistication, electrical safety in your physical plant will always be a complicated and important issue warranting its own comprehensive program. How you set yours up, of course, is entirely up to you, but might I suggest...

- addressing the topic of electrical safety with an ardent zeal

- developing a set of program objectives

- including the proper people in the program's construction, institution and application

- establishing an electrical safety committee comprised of electrically savvy individuals, working from the dictates of a committee charter similar to the safety charter discussed earlier.

- schedule regular inspections and meetings to cover pertinent issues

- budget sufficient monies for training and safety awareness programs

- establish equipment acquisition inspection and testing policies

- arrange for periodic checkout of existing equipment and systems per manufacturer's recommendations

- document all electrical related activities in the physical plant

Maintenance of High-Voltage Substations

When work is required on low-voltage electrical equipment you have two choices: work it live or work it dead. When work is required on a high-voltage substation, you also have two choices: work it dead or work it ...*dead*. Unless your electricians are former power company employees, specially trained and practiced in the use of electrical protective gear (0-500 volts with rubber goods and hot sticks for any voltage), I'd leave the high-voltage work for the utility workers. And believe me, considering the grief you could cause them with a mishap in your feeder yard, they feel the same way. Even if you're only visually inspecting them, it's a good idea to abide by OSHA's recommended general clearances of:

- 10 feet — 12,000 to 34,500 volts
- 11 feet — 69,000 volts (69 KV)
- 12'4" — 115,000 volts (115 KV)
- 16 feet —230,000 volts (230 KV)
- 20 feet — 345,000 volts (345 KV)

For informational purposes only, OSHA's recommended work clearances for qualified personnel for the same circuit capacities are:

- 2 feet — 12,000 to 13,500 volts
- 2'4" — 23,000 to 34,500 volts
- 3 feet — 69,000 volts (69 KV)
- 3'4" — 115,000 volts (115 KV)
- 5 feet — 230,000 volts (230 KV)
- 7 feet — 345,000 volts (345 KV)

(REMEMBER THE ANALOGY OF THE SNAKE)

To work safely on de-energized electrical equipment, the circuit must be isolated, through a disconnect switch that's been declutched (motor operated) or had the fuses pulled or a circuit breaker drawn out to form a visible air break. The circuits and equipment must then be tagged, locked out, tested dead and grounded. An envelope of protection around the work area must then be secured using colored tape and flags to form the perimeter. When the work is completed, the work area must be thoroughly inspected, a check made to assure the workmen are clear, the grounds removed and a second inspection made before re-energizing.

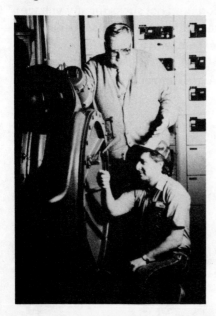

Figure 7-2. Did you remember to lock out the starter panels behind you? (*Courtesy of Johnson Controls*)

CARE OF PANELBOARDS

More in the realm of the certified electrician you'd find working for one of us is the care and maintenance of the building's many panel boards, fed by the substations we just discussed. Electrical panelboards are often neglected, though important major components in our electrical supply system on which regularly scheduled preventive maintenance should be performed along with the step-down transformers they are connected to. The following are general guidelines for maintaining panelboards rated at 600 volts or lower. Again, these units should not be worked on by anyone other than qualified personnel adhering to the safety dictates of the electrical profession, in servicing this type and capacity equipment.

Bear in mind, the conveyance of this information is not intended for use in performing the work, but rather to provide insight to the Evaluator regarding the nature and degree of work involved in caring for such equipment.

IMPORTANT! BEFORE PERFORMING ANY OF THE FOLLOWING OPERATIONS, TURN OFF ALL POWER SUPPLYING THE PANELBOARD AND CHECK THE VOLTAGE OF ALL INCOMING LINE TERMINALS TO POSITIVELY ASCERTAIN THAT THE EQUIPMENT IS TOTALLY DE-ENERGIZED.

1. A panelboard which has been carrying its regular load for at least 3 hours just prior to inspection should be field tested by feeling the deadfront surfaces of circuit breakers, switches, interior trims, doors and cabinet sides with the palm of the hand. If the temperature of these surfaces does not permit you to maintain contact for at least 3 seconds, this may be an indication of trouble and investigation is necessary.

2. Inspect the panelboard once each year or after any severe electrical fault.

   a. Look for any moisture or signs of previous wetness or dripping inside the panelboard. Condensation in conduits or dripping from outside sources is a common cause of panelboard failure.

   - Seal off any conduits which have dripped condensate, and provide a means for the conduit to drain.

   - Seal off any cracks or openings which have allowed moisture to come in from outside the enclosure. Where possible, eliminate the source of any dripping on the enclosure or other source of moisture.

   - Replace or thoroughly dry and clean any insulating material which is damp or wet or shows an accumulation of deposited material from previous wettings.

   b. If there is an appreciable accumulation of dust and dirt, clean out the panelboard by using a brush, vacuum cleaner, or clean lint-free rags. Avoid blowing dirt into circuit breakers or other equipment.

c. Carefully inspect all visible electrical joints and terminals

- Retighten bolts and nuts at bus joints if there is any sign of overheating or looseness. If joints appear to be badly discolored, corroded, or pitted, the parts should be disassembled or replaced or cleaned.

- Examine all wire or cable connections for evidence of looseness or overheating. Retorque if necessary. If major discoloration or cable damage is apparent, replace the damaged parts and remove the damaged portion of the cable.

- Closely examine fuse clips. If there is any sign of overheating or looseness, check spring pressure, tightness of clamps, etc. Replace the fuse clips if the spring pressure compares unfavorably with that of other identical fuse clips on the panelboard.

- Look for signs of deterioration in insulating material or melting of sealing wax. Replace such insulating parts, and assemblies where sealing was has melted. Be sure the condition which caused the overheating has been corrected.

d. Check operation of all mechanical components

- Exercise switch operating mechanisms and external operators for circuit breakers, making sure that they quickly and surely throw the contacts fully on and fully off

- Check the integrity of all electrical and mechanical interlocks and padlocking mechanisms

- Wherever possible, check all devices for missing or broken parts, proper spring tension, free movement, rusting or corrosion, dirt and excessive wear

- Adjust, clean and lubricate or replace parts as required

Electrical Safety

e. Examine all readily accessible arc chutes and insulating parts for cracks or breakage and for arc spatter, sooty deposits, oil or tracking.

- clean off arc spatter, oil, and sooty deposits, but replace parts if appreciable material has burned away or if material is charred or tracked

- replace insulating parts and arc chutes that are cracked or broken

3. Clean and dress copper electrical contacts, blades and jaws when inspection indicates the need to do so. Avoid removing metal from silver contact surfaces.

   a. Dress sliding contacts to remove projecting burrs or arc transferred material which interferes with free movement

   b. Clean contact surfaces to remove black oxides. Use fine aluminum oxide paper and remove as little material as possible. Keep metal abrasive particles entirely out of the panelboard.

   c. Replace blades or jaws which are burned away enough to appreciably change their shape or to interfere with the tapered edges, etc., which allow them to act freely.

   d. Where contacts show signs of overheating, compare spring pressures and stiffness of parts with those of identical good contacts. Replace springs, jaws, blades, or contacts if they have been softened or otherwise damaged by heat.

4. Lubricate the operating parts of switch mechanisms, etc., according to the manufacturer's instructions which are usually printed on diagrams or labels.

   a. Use clean, nonmetallic, light grease or oil as instructed.

   b. Do not oil or grease parts of molded case circuit breakers.

c. If no instructions are given on the devices, sliding copper contacts, operating mechanisms and interlocks may be lubricated with clean, light grease.

d. Wipe off excess lubrication to avoid catching dirt.

5. Operate each switch or circuit breaker several times to make sure that all mechanisms are free and in proper working order.

6. Check insulation resistance:

    a. If a severe electrical fault has occurred.

    b. If it has been necessary to replace parts or clean insulating surfaces.

    c. If the panelboard has been exposed high humidity, condensation or dripping moisture.

7. Check fuses to make sure they have the proper ampere rating and interrupting rating. Make sure that noncurrent-limiting fuses are never used as replacements for current-limiting fuses. Never attempt to defeat rejection mechanisms which are provided to prevent the installation of the wrong type of fuses.

Systems Maintenance

A building's electrical system can be compared to a person's circulatory system, in that the electricity it supplies is the life blood that keeps the building and its vital parts operational. And, like your own kidneys, heart and arteries, neglecting to maintain the system's components and wiring can lead to malfunctions and the eventual cessation of service. This is not a pretty picture!

Aside from assuring its continued operation, regularly scheduled preventive maintenance performed on a building's electrical systems provides a safer environment for employees working with or near it, reduces repair costs and production losses and uncovers defects that might otherwise go undetected until after the accident occurs. Paying attention to and properly caring for your system can improve its dependability and extend its life expectancy. A fairly comprehensive program can be had for a relatively minor investment; the expense of maintaining it is usually less than the cost of replacing it. An annual check of the tightness of electrical connections, the dielectric strength of insulation, the functionality of moving parts and the integrity of the system's protective devices is all that's called for, most of which can be done in-house by your own people. This simple program can then be supplemented by an in-depth thermographic analysis conducted every 3 to 5 years. What to check? Minimally...

- Transformers
- Bus and bus duct
- Panelboards
- Circuit breakers
- Fuse clips
- Relays and Meters
- Protective instrumentation
- Power cables
- Motor starters
- Batteries and chargers
- ... etc.

GROUND FAULT INTERRUPTION

We're all knowledgeable as to the whys and wherefores of the resettable, ground fault circuit interrupters (GFI's) required by the National Electrical Code in wet areas for personal protection, but have you ever considered the need, or for that matter the existence of ground fault protection in your facility's main electrical feed?

Ground fault protection equipment detect ground fault currents and provide system tripping through a disconnect prior to the occurrence of serious damage from otherwise energized circuits downstream.

Its three basic components are: a sensing transformer, a relay and a switching device which is tripped by the relay. Under normal conditions, current supplied through a transformer or bus to a system, returns to the source through the phase and neutral conductors. Current returning by way of an alternate path, as through the ground or grounding conductor, indicates a phase-to-ground leak that, while possibly too small to trip the system's overcurrent protectors, can generate high temperatures resulting in damage to equipment in short order. In designing ground fault protection, the selection of trip settings and sensor time-delays is extremely critical to avoiding nuisance episodes after installation. If you aren't one yourself, you might want to bring in an EE (Electrical Engineer) to help you with this one.

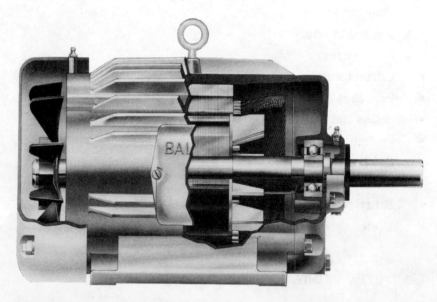

Figure 7-3. Single phase conditions can wreak havoc with unprotected three-phase motors. *Photo courtesy of Baldor Electric Company*

SINGLE PHASE PROTECTION

Single phase is an unbalanced, overcurrent condition in a three-phase electrical circuit which exists when one of the phases opens. It can occur on either the primary or secondary side of a distribution transformer and cause havoc with unprotected three-phase motors, lighting systems

Electrical Safety

and control devices.

Primary side single phasing can be caused by broken wires due to inclement weather, lightning, fallen trees or a number of other reasons not under the control of the electrical consumer. It causes unbalanced voltages on motor circuits which consequently subjects them to unbalanced currents. The amount of current in each phase depends on the type of transformer and other parameters. In secondary side opens, the current in the two energized legs will normally increase to 173 percent factor and, where the motors have high inertia loads, currents can reach locked rotor values. If each leg is protected with a thermal overload relay, the current overload will be sensed and the motors protected. But again, I defer to our electrical cousin's expertise in dealing with the problem.

EMERGENCY POWER GENERATION

Should Mr. Murphy pay you a visit before you're able to get your electrical house in order, your last bastion of defense (unless you're fortunate enough to be fed by two separate substations) is your emergency power installation. If you don't have one, install one; if you do, maintain it... well! Because, when the chips (wires) are down, it will take care of you only as well as you took care of it. Autonomous power pays dividends in improved safety, comfort, and security and provides potential for peak shaving and sale of power back to your electric utility. While you're giving some thought to the subject, take a look at this generator testing and maintenance policy (pages 95-103).

SAFETY TIPS

Far from exhausting the subject of electrical safety (but then I intended only to touch on the topic), let me leave you with these electrical pearls of enlightenment:

- Distribution lines are overhead and underground—notify your utility company prior to using a crane or digging.

- If the power goes off, find out why before taking any action, then repair the trouble before re-energizing circuits.

- Keep circuits from becoming overloaded or unbalanced.

- Schedule portable power tools for frequent maintenance and allow only the use of double insulated tools and/or those with three-prong grounded plugs.

- Locate ladders, antennas and poles away from incoming power lines.

- Use only tools, lights, fixtures and extension cords designated for exterior use outside, and plug them only into GFI protected circuits.

- Place Class "C" fire extinguishers within easy reach, in areas where electrical fires are a possibility.

- If you smell smoke or see flame, unplug the device involved or turn off the power at the main control panel.

- Never touch anyone in contact with a power source—turn off the power! If you can't turn off the power, separate the victim from the power source with a non-conductor.

- Never allow electrical rooms to be used for storage of any kind.

Figure 7-4. Emergency Generator Testing Policy

```
                              Policy Number    _____
                              Effective Date   _____
                              Revision Number  _____
                              Page             _____
```

I.   ORIGINATOR: Facilities Manager

II. TITLE: Emergency Generator Testing & Maintenance

III. SCOPE: All Departments

IV. PURPOSE: To ensure the operation of the companys'
                   autonomous electrical power generators during
                   interuptions of normal supply and to alert all
                   departments to the equipments scheduled test
                   periods.

V.   SCHEDULE:

                   The generators will be tested, under full load
                   conditions, on a permanent basis, the first
                   Wednesday of each month between 5:30 AM and 6:00
                   AM.

NOTE: For the protection of the operator and in anticipation of problems
that might arise -- "NO LESS THAN TWO MAINTENANCE PERSONNEL MUST BE PRESENT
FOR THE DURATION OF THE TEST."

VII. PROCEDURE:

                   Testing personnel must report to the maintenance
                   shop no later than 5:00 AM and must follow these
                   procedures in the order that they are listed.

Pre-test activity

1. Check the diesel-generators, main electrical switchgear and mechanical
    equipment rooms for obvious discrepancies.

Figure 7-4. Emergency Generator Testing Policy (cont'd)

```
                                    Policy Number    _____
                                    Effective Date   _____
                                    Revision Number  _____
                                    Page             _____
```

2. While one of the testers makes a pre run check of the equipment, according to the posted start-up/shut down procedures, the second is to notify the following that a test is imminent;

   a) Fire Department 458-5200
   b) Electric Utility 458-3600
   c) Switchboard . . . . -- 0 --
   d) Security . . . . Ext 911
   e) Night Supervisor . . . . -- 0 -- (for page)
   f) Power House Engineer . . . . Ext 609

NOTE: If any operational problems occur during the course of the pre start checks or if the possibility of adverse service interruption becomes apparent during telephone notifications, the test is to be aborted for reschedule at a later date.

Testing

1. Beginning at 5:30 Am, the guidelines set forth in the startup/shut down procedures for POWER UP sequencing are to be strictly adhered to.

2. All operating checks are to be made and recorded on the generator operating log sheet during the test, as outlined in the standard operating procedures manual.

3. At 6:00 AM the POWER DOWN sequence procedures are to be strictly adhered to.

Post test activity

1. After normal electrical power has been restored, one of the testers will stand by the generator set as it idles down, while the other notifies all, that the test has been completed.

2. A copy of the generator operating log sheet is to be put on the desk of the Facilities Manager.

Electrical Safety

Figure 7-4. Emergency Generator Testing Policy (cont'd)

```
                                    Policy Number    _____
                                    Effective Date   _____
                                    Revision Number  _____
                                    Page             _____
```

VII.   OPERATION:

PRE-START CHECKLIST

1. Place the control panel selector switch which is labeled "Manual-Off-Auto" in the "Off" position.

2. Check generator and electrical rooms for any obvious discrepancies.

3. Check that engine and radiator are free of debris, foreign objects, and loose or broken parts.

4. Check fuel level in day tank.

5. Check governor oil level (Woodward governor).

6. Check engine oil level with the dipstick.

7. Check coolant level; it should be above the baffle plate, one-half inch below the fill pipe.

8. Check fan and alternator for wear and tension; deflection should be 9/16 to 13/16 inches @ 25 lbs. force.

9. Check the oil heater to insure it is operating.

10. Check to insure that the engine, generator, radiator and intake and exhaust louvers are free of foreign objects and debris.

11. Examine the motor operated louvers to insure that their linkage is in proper adjustments so as to keep the louvers tightly closed when the generator set is at standstill.

12. Make certain that all fuel oil valves are in the "open" position.

13. Check the battery water level and measure the specific gravity of each cell. Inspect the battery terminals for signs of corrosion and tightness of connectors. Clean the top of the battery.

Figure 7-4. Emergency Generator Testing Policy (cont'd)

```
                            Policy Number     _____
                            Effective Date    _____
                            Revision Number   _____
                            Page              _____
```

14. Correct any discrepancies before starting the engine.

15. Place the control panel selector switch which is labeled "Manual-Off-Auto" in the "Auto" position.

16. Make sure the generator breaker is in the "On" position.

POWER UP SEQUENCE

1. Pull the main electrical switch into the "Off" position and allow the generator time to accept the load.

2. Pull the EC main switch into the "Off" position and allow the generator time to accept the load.

3. Pull the EQE main switch into the "Off" position and allow the generator time to accept the load.

NOTE: Load acceptance by the generator should be checked at the generator control panel after each main switch is pulled. Loading of the generator is accomplished when an increase in amperes is manifested on the gauge. If transfer does not take place, through the transfer switch, for each main circuit, the test should be aborted and the Facilities Manager should be contacted immediately.

OPERATING CHECKLIST

1. Stay with the generator while it is in the operating mode, being cognizant of any unusual noises or abnormal vibration of the unit which would cause you to abort the test.

2. Using the operating log, record all required readings as you observe the units operation.

Figure 7-4. Emergency Generator Testing Policy (cont'd)

```
                          Policy Number    _____
                          Effective Date   _____
                          Revision Number  _____
                          Page             _____
```

POWER DOWN SEQUENCE

Return the four main switches to the "On" position in the reverse order that they were deenergized and observe transfer switches for proper operation.

POST TEST PROCEDURE

1. Stay with the generator during its idle down period until it shuts down completely.

2. Check to insure that the control panel switch labeled "Manual-Off-Auto" is in the "Auto" position.

3. Check the fuel level in the fuel tank and refill if necessary.

4. Inspect for proper closing of the intake and exhaust louvers.

5. Record any unusual or abnormal circumstances or situations that arose during the course of the test, on the operating log, under the "Remarks" section.

6. Place the completed operating log on the Managers desk.

VIII.   MAINTENANCE:

Once testing has concluded, the operators will perform the following minimum preventive maintenance procedures on the diesel generators, indicated as being due for completion in their work journals ---

Figure 7-4. Emergency Generator Testing Policy (cont'd)

Policy Number	_____
Effective Date	_____
Revision Number	_____
Page	_____

Daily

1. Check generator and electrical rooms for obvious discrepancies.

2. Check that engine and radiator are free of debris, foreign objects and loose or broken parts.

3. Check that louvers are in the closed position.

4. Check that the generator breaker is in the "on" position and the panel selector is in the "auto" position.

5. Check the oil heater to insure it is operating.

Weekly

1. Check the battery level and measure the specific gravity of each cell. Inspect the battery terminals for signs of corrosion and tightness of connectors. Clean the top of the battery.

2. Check for proper tension of the belts.

Monthly

1. Engine Oil Level – observe dipstick with engine idling (oil hot), maintain oil level between the add and full marks on dipstick.

2. Oil Filter Change Indicator – observe the red plunger with engine running at full load. If the plunger fills half the plastic window, install a new oil filter element.

3. Air Cleaner Indicator – observe the position of the indicator piston. If piston locks in the "UP" position, service the element.

4. Governor – maintain oil level between marks on sight glass.

Electrical Safety

Figure 7-4. Emergency Generator Testing Policy (cont'd)

```
                              Policy Number    _____
                              Effective Date   _____
                              Revision Number  _____
                              Page             _____
```

5. Coolant Level – observe level with engine stopped and cold, maintain level in range above baffle plate and ½" below fill pipe.

6. Fuel Filter Housing – drain water and sediment

7. Fan/Alternator Belts – check wear and belt tension, belt should deflect 9/16" to 13/16" @ 25 lbs. force.

8. Battery Electrolyte Level – maintain electrolyte level to the base of each vent well.

Quarterly

1. Engine Breather – clean, use clean solvent

2. Water Cooling System – add cooling system inhibitor

3. Radiator Fan – lubricate filling – 2 strokes.

Semi-Annually

1. Oil Filter – change oil filter

2. Governor – change oil, fill to full mark on sight gauge

3. Governor Motor – lubricate wick in oil cup

4. Shutoff Control/Reset Level – lubricate one fitting – 2 strokes, inspect cable for kinks and movement drag.

5. Tachometer Drive – lubricate one fitting – 1 stroke

Annually

1. Valve Lash – Check adjustment: Inlet valve – .012" (6 clicks from zero lash) --- Exhaust Valve – .026" (13 clicks from zero lash).

Figure 7-4. Emergency Generator Testing Policy (cont'd)

```
Policy Number    _____
Effective Date   _____
Revision Number  _____
Page             _____
```

2. Valve Rotators - observe rotation of valves with engine idling

3. Electric Set Generator - remove 2 plugs, install upper fitting, lubricate until old grease appears.

4. Remove, clean and re-install battery connections

5. Change crank case oil and oil filter.

6. Change fuel filter.

7. Inspect air filter and change if necessary.

8. Drain sufficient fuel from bottom of day tank to eliminate all water and sediment.

9. Vacuum clean generator and exciter.

10. Wipe down engine.

11. Clean and lubricate all linkages and grease fittings.

<u>Bi-Annually</u>

1. Drain coolant and replace all hoses.

2. Refill cooling system with 50% ethylene glycol solution, 2 quarts of cooling system conditioner and distilled water and wire new date tag to radiator.

3. Replace all drive belts.

4. Tune and adjust the engine.

5. Check exhaust the system for leaks.

Figure 7-4. Emergency Generator Testing Policy (conclusion)

```
Policy Number    _____
Effective Date   _____
Revision Number  _____
Page             _____
```

_____

Facilities Manager   Date

_____

Safety Chairman   Date

_____

Administrative   Date
Review

CHAPTER 8
ENVIRONMENTAL COMPLIANCE

In 1949 the British writer, Eric Arthur Blair (George Orwell), wrote a novel (*1984*) satirizing life under the constant surveillance of "Big Brother." Those of us familiar with the work swear that we've been suffering under that imperative every year ever since. "Big Brother" is alive, aware and just around the corner—taking many forms. He can be as big as HUD (U.S. Department of Housing and Urban Development) or as small as the town housing inspector, but more often than not he's a governing body comprised of a panel of experts who decide our fate regarding public policy—with or without our input.

Whatever his stature and whether he's been voted in, appointed, hired, legislated or has simply assumed the position, I guarantee you he's about and watching. Just so you'll be able to recognize him when you see him, here are a few of the disguises he uses when dealing with Facilities Managers. The following is a listing of major Federal laws covering the environment; at the state level, I'll use New York regulations as being representative of all states, because that's where I live... I'm sure you have or will soon be subjected to comparable legislation where you reside. Now... where to begin? How about...

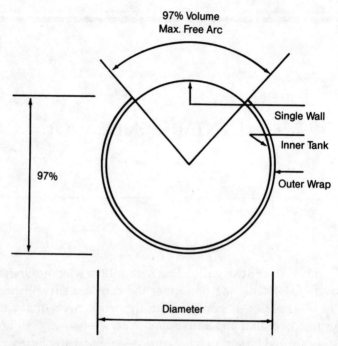

Figure 8-1. Now if they'd just do the same thing with the supertanksers that deliver it... (common dual wall design with 300-degree outer tank wrap).

PETROLEUM STORAGE LAWS

State Equivalency: New York Oil Spill Prevention, Control and Compensation Act

Intent: Prevention of petroleum discharges to lands and waters through improved safeguards in storage and handling, prompt cleanup and removal

Requires: Facility registration, operating licenses for major facilities, leak detection, corrective action, technical standards

Coverage: Petroleum bulk storage facilities storing more than 1100 and less than 400,000 gallons, (MPSF) major petroleum storage facilities storing more than 400,000 gallons

Funding: Registration fees and state appropriations

Inspection: DEC (Department of Environmental Conservation) may inspect facilities and records

Data: Facilities must maintain measurement and inventory records

Application: Underground and above ground tanks used for storing and supplying fuel for boilers (heating process), diesel engines (electrical power generation) and fleet vehicles.

Figure 8-2. Sample MSDS

Material Safety Data Sheet

QUICK IDENTIFIER
Common Name: (used on label and list)

May be used to comply with OSHA's Hazard Communication Standard.
29CFR 1910. 1200. Standard must be consulted for specific requirements.

SECTION 1 -

Manufacturer's Name

Address

City, State, and ZIP

Signature of Person
Responsible for Preparation (Optional)

Emergency Telephone No.

Other Information Calls

Date Prepared

SECTION 2 – HAZARDOUS INGREDIENTS/IDENTITY

Hazardous Component(s) (chemical & common name(s))	OSHA PEL	ACGIH TLV	Other Exposure Limits	(optional)	CAS NO.

SECTION 3 - PHYSICAL & CHEMICAL CHARACTERISTICS

Boiling Point		Specific Gravity (H$_2$O=1)		Vapor Pressure (mm Hg)
	Vapor Density (Air = 1)			
Solubility in Water		Reactivity in Water		
Appearance and Odor		Melting Point		

SECTION 4 – FIRE & EXPLOSION DATA

Flash Point F. C.	Method Used	Flammable Limits in Air % by Volume	LEL Lower	UEL Upper
Auto-Ignition Temperature	Extinguisher Media			
Special Fire Fighting Procedures				

Unusual Fire and Explosion Hazards

Figure 8-2. Sample MSDS (cont'd)

SECTION 5 – PHYSICAL HAZARDS (REACTIVITY DATA)

Stability   Unstable ☐   Conditions
            Stable   ☐   to Avoid

Incompatability
(Materials to Avoid)

Hazardous
Decomposition Products

Hazardous   May Occur   ☐   Conditions
Polymerization   Will Not Occur ☐ to Avoid

SECTION 6 – HEALTH HAZARDS

1. Acute   2. Chronic

Signs and
Symptoms of Exposure

Medical Conditions Generally
Aggravated by Exposure

| Chemical Listed as Carcinogen or Potential Carcinogen | National Toxicology Program | Yes ☐ No ☐ | I.A.R.C. Monographs | Yes ☐ No ☐ | OSHA | Yes ☐ No ☐ |

Emergency and
First Aid Procedures

ROUTES OF ENTRY
1. Inhalation
2. Eyes
3. Skin
4. Ingestion

SECTION 7 – SPECIAL PRECAUTIONS AND SPILL/LEAK PROCEDURES

Precautions to be Taken
in Handling and Storage

Other
Precautions

Steps to be Taken in Case
Material is Released or Spilled

Waste Disposal
Methods (Consult federal, state, and local regulations)

SECTION 8 – SPECIAL PROTECTION INFORMATION/CONTROL MEASURES

Respiratory Protection
(Specify Type)

| Ventilation | Local Exhaust | Mechanical (General) | Special | Other |

| Protective Gloves | | Eye Protection | | |

Other Protective
Clothing or Equipment

Work/Hygienic Practices

IMPORTANT
Do not leave any blank spaces. If required information is unavailable, unknown, or does not apply, so indicate.

WORKER RIGHT-TO-KNOW

Federal law:	Occupational Safety and Health Act; Hazard Communications Standard
State Equivalency:	New York Worker Safety Regulations
Intent:	Identification, listing and disclosure of hazardous substances; employee training and education
Requires:	OSHA labeling information; access to employee health/exposure records
Coverage:	Workplaces with hazardous chemical substances
Funding:	Federal budget, state appropriations, grants for education and training services
Inspection:	OSHA/Department of Labor may inspect facilities and employee health records
Data:	OSHA and the New York State Department of Labor both supply hazardous substance information to employees
Application:	Chemicals, solvents and hazardous substances used by or near employees throughout the workplace

Environmental Compliance

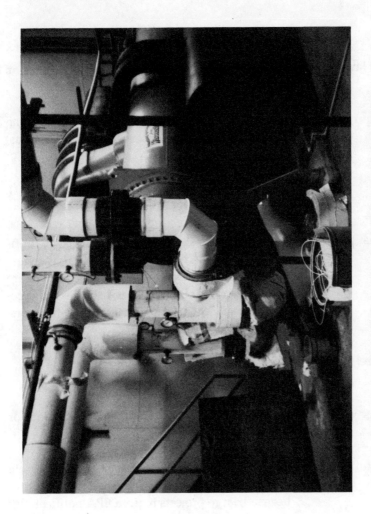

Figure 8-3. Fiberglass insulation is more difficult to apply than asbestos lagging; but the effort is well worth it...

ASBESTOS

Federal law: OSHA and EPA (Environmental Protection Agency) Asbestos Laws

State Equivalency: New York State Department of Labor, Education and Environmental Conservation (air emissions) Asbestos Laws

Intent: Regulation of asbestos encapsulation or removal, worker safety; survey of asbestos in schools.

Requires: Regulation of asbestos manufacturing, identification, handling, removal and disposal; management of asbestos abatement in schools, child-care facilities and public buildings; asbestos management plans; licensing and permitting of contractors, workers and job sites.

Coverage: Asbestos and ACM (Asbestos Containing Material)

Funding: EPA grants, fees and applications, State grants for school projects and fines

Inspection: Contractors and their employees must complete EPA certified courses and be state certified and licensed; large projects require EPA notification.

Data: Project records must be maintained (40 years in New York State)

Application: Steam pipe lagging, boiler and water tank insulation, floor and ceiling tile, refractory materials, etc.

Environmental Compliance

Figure 8-4. No longer can we just "dump it out back"!

HAZARDOUS WASTE

Federal law:	RCRA (U.S. Resource Conservation and Recovery Act), Hazardous and Solid Waste Amendments (HSWA)
State Equivalency:	New York Solid and Hazardous Waste Management Laws
Intent:	Regulation of Hazardous Waste generators, transporters and TSD (treatment, storage and disposal facilities; assurance of safe management of hazardous waste from "cradle to grave."
Requires:	Permits, technical standards, corrective action for releases, transportation manifests, monitoring of groundwater, record keeping and labeling
Coverage:	Federal and state appropriations, permit and license fees, fines and penalties
Inspection:	DEC inspects permitted TSD facilities, EPA identification number required
Data:	Manifests must be maintained by generators and transporters; TSD facilities retain records
Applications:	Disposal of everything from hazardous chemicals to sharp objects

Environmental Compliance

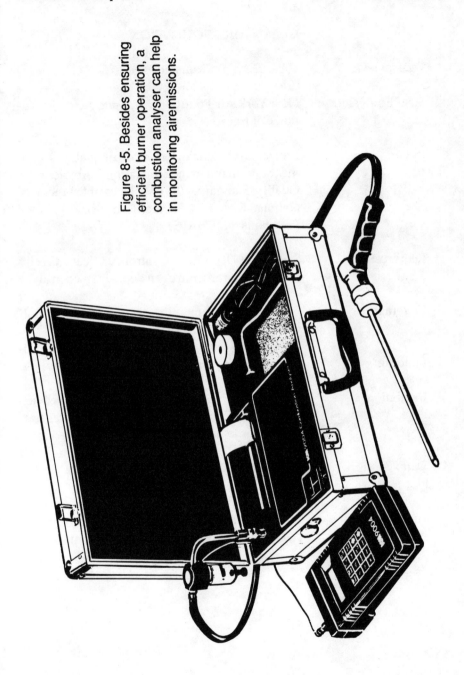

Figure 8-5. Besides ensuring efficient burner operation, a combustion analyser can help in monitoring air emissions.

CLEAN AIR REGULATION

Federal law: CAA (Federal Clean Air Act)

State Equivalency: New York Air Pollution Act, State Acid Deposition Control Act

Intent: Protection and enhancement of air quality; attainment and maintenance of National Ambient Air Quality Standards (NAAQS); establishment of National Emissions Standards for Hazardous Air Pollutants (NESHAPS).

Requires: State Acid Deposition Control Act deals directly with sulphur and nitrogen deposition control

Coverage: Sources of air contaminants; hazardous air pollutants and volatile organics

Funding: EPA grants, general revenues, fine and penalties

Inspection: Permits and operating certificates required for sources; DEC may inspect; NY law requires trial burn prior to burning of hazardous waste

Data: DEC submits an annual emissions report to the EPA

Application: Emissions from kiln, boiler and incinerator smoke stacks; motor vehicles and process exhausts.

Typical Partial Analyses Of Some City Water Supplies

City		Los Angeles		Houston		Denver		Chicago	Miami		New York		
Constituent	Source (ppm as)	Weymouth Treatment Plant	San Fernando Reservoir	San Jacinto Plant	East End Wells	Moffat Plant	Marston Lake Plant	South District Plant	Hialeah Plant	Orr Plant	Catskill Supply	Croton Supply	Jamaica Wells
Total Hardness	$CaCO_3$	200	81	59	14	34	99	129	75	58	18	54	191
Calcium	$CaCO_3$	130	60	53	10	25	65	85	55	50	14	35	113
Sodium	Na	151	32	11	177	3	24	4	23	13	2	3	17
Potassium	K	4	4	2	1	0.4	2	1.2	-	-	0.7	1.3	1.6
Total Alkalinity	$CaCO_3$	114	108	21	307	23	66	101	26	37	6	30	130
Chloride	Cl	88	16	27	59	0	32	10	42	16	4	9	28
Sulfate	SO_4	285	23	26	8	19	38	25	26	24	12	19	53
Fluoride	F	0.4	0.5	0.3	1.5	0.8	0.9	0.9	1	1	0	0.1	0
Nitrate	NO_3	1.2	0.2	0	0	0.4	0.7	0.4	0.3	0.3	0.3	0.3	8.6
Silica	SiO_2	9	31	6	16	8	6	1.7	9	7	3	5	21
Total Dissolved Solids	--	678	212	129	456	39	175	165	185	110	42	97	283
pH	--	8.3	8.1	6.5	7.7	7.2	7.4	7.7	9.0	8.7	6.7	7.5	7.1

Figure 8-6. Don't those LA waters look inviting?

WATER QUALITY

Federal law:	Federal Clean Water Act
State Equivalency:	New York Water pollution control
Intent:	Restoration, enhancement and maintenance of surface and ground water quality; prevention of waste discharges that threaten public health and the environment
Requires:	Classification and deregulation of waterway use and development of water quality plans by the DEC; the setting of discharge standards by the EPA and DEC; issuance of permits by the DEC for discharging into water
Coverage:	Restricts discharge of pollutants into water; regulates sources of discharge; DEC imposed monitoring requirements
Funding:	Federal grants, permit fees and penalties
Inspection:	DEC issues permits, inspects facilities and records
Data:	Facilities must submit monitoring data; DEC implements water quality surveillance to monitor ground and surface water and publishes its findings
Application:	On site sewage treatment, effluent discharge into municipal sewage system, cooling tower and boiler water treatment, combination storm sewers

WATER ANALYSIS

		Parts per Million	Parts per 100,000	Grains per U.S. Gallon	Grains per Imp. Gallon
1 part per million	=	1.0	0.10	0.058	0.07
1 part per 100,000	=	10.0	1.00	0.585	0.70
1 grain per U.S. Gal.	=	17.1	1.71	1.000	1.20
1 grain per Imp. Gal.	=	14.3	1.43	0.833	1.00

HARDNESS AS $CaCO_3$

		Parts per Million	Grains per U.S. Gallon	Clark Degrees	French Degrees	German Degrees
1 part per million	=	1.0	0.058	0.07	0.10	0.056
1 grain per U.S. Gal.	=	17.1	1.000	1.20	1.71	0.958
1 Clark degree	=	14.3	0.833	1.00	1.43	0.800
1 French degree	=	10.0	0.583	0.70	1.00	0.560
1 German degree	=	17.9	1.044	1.24	1.78	1.000

Water at 39.2 F (4 C),
maximum density

1 cubic foot	= 62.4 lb
1 cubic meter	= 1000 kg
1 pound	= 0.01602 cu ft
1 liter	= 1.0 kg

Water at 62 F (16.7 C)

1 cubic foot	= 62.3 lb
1 pound	= 0.01604 cu ft
1 U.S. gallon	= 8.34 lb

Figure 8-7. …just thought you'd want to know.

SAFE DRINKING WATER

Federal Law:	Federal Safe Drinking Water Act
State Equivalency:	New York Regulation of Public Water Supplies
Intent:	Assure the quality of drinking water
Requires:	Regulation of construction, monitoring, testing, and standards; operation of drinking water systems
Coverage:	Hazardous contaminants; testing and standards; operation of drinking water systems
Funding:	Federal grants and state appropriations
Inspection:	Permits required for operation of public water systems; Department of Health approves applications by realty subdivisions for installation of water supplies and sewage systems
Data:	Operation, maintenance and monitoring reports are required to be maintained.
Application:	Incoming water lines, water heaters and coolers, ice machines

Environmental Compliance

Figure 8-8. At least we can keep the *little* pests at bay.

INSECTICIDES

Here's a roundup of common chemicals used to control insects. The chemical names are generic.

This information is provided not only to help you set up an application schedule, but to give you an idea of how long to take precautions. You'd be wise to avoid contact with treated areas for the active life of the chemical.

The following chemicals have been banned: aldrin, chlordane, DDT, dieldrin, heptachlor, kepone, lindane, mirex. Any product containing these chemicals is dangerous to humans and animals and *should not* be used, even if you can still find it for sale.

Chemical	Use it to control	Active life*
allethrin	insects	short
baygon	crawling insects	long
carbaryl	insects	medium
chlorobenziate	spiders and mites	medium
coumaphos	rodents	mdeium
DDD	insects	long
diazinon	insects	long
dichlovos	parasites	short
dicofol	mites	long
dimethoate	vegetable pests	long
dioxathion	insects	long
endosulfan	insects	medium
fenthion	mosquitoes and flies	medium
malathion	insects	medium
metaldehyde	slugs and snails	long
methoxychlor	insects	medium
naled	insects	short
oils	insects	long
ovex	mites	long
pyrethrins	insects	short
ronnel	parasites	medium
sulfur	mites	long
toxaphene	insects	long
trichlorfon	insects	long

*A short active life is considered to be three days; a medium active life, fourteen days; a long active life, thirty-five days.
NOTE: Many of the products listed are of the broad-spectrum type; a listing of "insects" under the heading "Use it to control" means this chemical can be used on a wide variety of insects. Check the package label.

PESTICIDES

Federal Law:	FIFRA (Federal Insecticide, Fungicide and Rodenticide Act)
State Equivalency:	New York Pesticide Regulation
Intent:	Regulation of pesticides to prevent adverse health and environmental effects
Requires:	Registration and labeling of pesticides; regulation of storage, use, transportation, disposal and clean up; licensing of dealers and applicators
Coverage:	General, restricted and mixed use pesticides
Funding:	Federal and state funding; annual registration and licensing fees
Inspection:	Registration of pesticides, dealers, operators and applicators; NY law requires a purchasing permit
Data:	The EPA has a national monitoring plan; dealers and applicators must maintain records of pesticide use
Application:	Pest and vermin control

Figure 8-9. ...now we're even gift wrapping our garbage.

Environmental Compliance

TOXIC CLEAN UP

Federal Law: CERCLA (The Comprehensive Environmental Response and Liquidity Act), TSCA (Toxic Substance Control Act)

State Equivalency: Inactive Hazardous Waste Sites Law; Environmental Quality Bond Act of 1986

Intent: Location, assessment and cleaning of contaminated sites; reporting of hazardous chemical releases; prioritize inactive sites for clean up

Requires: DEC leadership role in clean up activities through registry of inactive sites; remedial programs for clean up of specific sites

Coverage: Releases of hazardous substances above reportable quantities; contaminated sites not currently holding a permit issues by federal and state authority

Funding: Responsible parties when identified; superfund; $500 million LUST fund; environmental quality bond act; hazardous waste remedial fund; collected fees and surcharges

Inspection: The EPA and DEC monitor clean up progress; DEC may inspect sites and records regarding past disposal practices

Data: Site specific data; clean up plan; registry maintained by the DEC

Application: PCB (Polychlorinated Biphenyl) contaminated electrical gear (transformers)

CHAPTER 9
SCORING THE EVALUATION

Now that you've had a chance to decide what you're going to look at and the extent to which you're going to look at it... let's get busy. Here's a scorecard I've devised to jog your memory and keep the evaluation proceeding on track. Whether the POE is to be performed by yourself and your subordinates, hired out, or conducted by way of some combination of the two, the information gathered should be evaluated and summarized here for later reference when reporting findings and detailing corrective action plans. Based on my past experiences, I've attempted to construct as comprehensive a checklist of considerations as I felt would be common to most Facilities Managers. For those of you with special needs, I left you some space at the end of each section to pen them in.

Before you begin, here are the scoring parameters again, to save you from looking them up.

- Risk Valuations
  - 0   Elements are correct, present & functioning
  - 1   Elements are correct, present but non-functional
  - 2   elements are incorrect or not present

- Risk Factors
  - H   Human
  - M   Mechanical
  - E   Environmental
  - P   Procedural

- Risk Exposure

(score)	(level)
0 - 2	low
3 - 5	meaningful
6 - 8	high

Good luck with your evaluation! I'll see you when it's finished to help sort out the information you will have compiled.

FACILITIES EVALUATION SCORECARD

I. SITE CHARACTERISTICS

Area/Item Evaluated (notation)	Risk Factor Evaluation H/M/E/P	Score/ Exposure Level

A. Existing Conditions

1. vehicular ingress & egress   / / /   /

(notations)_____
_____
_____
_____

2. handicap & pedestrian access   / / /   /

(notations)_____
_____
_____
_____

3. drives and walkways   / / /   /

(notations)_____
_____
_____
_____

Scoring the Evaluation

Area/Item Evaluated (notation)	Risk Factor Evaluation <u>H/M/E/P</u>	Score/ Exposure <u>Level</u>
4. pavement to dirt ratio	/ / /	/

(notations)_____
_____
_____
_____

| 5. parking layout/adequacy | / / / | / |

(notations)_____
_____
_____
_____

| 6. | / / / | / |

(notations)_____
_____
_____
_____

| 7. | / / / | / |

(notations)_____
_____
_____
_____

Area/Item Evaluated   Risk Factor   Score/
                             Evaluation   Exposure
(notation)   H/M/E/P   Level

8.   / / /   /

(notations)_____
_____
_____
_____
_____

9.   / / /   /

(notations)_____
_____
_____
_____

B. Grounds Maintenance

Area/Item Evaluated   Risk Factor   Score/
                             Evaluation   Exposure
(notation)   H/M/E/P   Level

1. equipment maintenance   / / /   /

(notations)_____
_____
_____
_____
_____

2. signage and lighting   / / /   /

(notations)_____
_____
_____
_____
_____

Scoring the Evaluation

Area/Item Evaluated	Risk Factor Evaluation	Score/ Exposure
(notation)	H/M/E/P	Level

3. fences and outbuildings   / / /   /

(notations)_____
_____
_____
_____

4. landscaping & groundskeeping   / / /   /

(notations)_____
_____
_____
_____

5. snow removal efforts   / / /   /

(notations)_____
_____
_____
_____

6.   / / /   /

(notations)_____
_____
_____
_____

Area/Item Evaluated   Risk Factor   Score/
                             Evaluation   Exposure
(notation)   H/M/E/P   Level

7.   / / /   /

(notations)_____

_____

_____

_____

8.   / / /   /

(notations)_____

_____

_____

_____

9.   / / /   /

(notations)_____

_____

_____

_____

II. BUILDING STRUCTURE

A. Interior

Area/Item Evaluated   Risk Factor   Score/
                             Evaluation   Exposure
(notation)   H/M/E/P   Level

1. stability of load-bearing walls   / / /   /

(notations)_____

_____

_____

_____

Scoring the Evaluation

Area/Item Evaluated (notation)	Risk Factor Evaluation H/M/E/P	Score/ Exposure Level
2. functional adjacencies	/ / /	/

(notations)_____
_____
_____
_____

3. compartmentalization/partitioning / / /   /

(notations)_____
_____
_____
_____

4. floor loading   / / /   /

(notations)_____
_____
_____
_____

5. floor, wall and ceiling systems   / / /   /

(notations)_____
_____
_____
_____

Area/Item Evaluated (notation)	Risk Factor Evaluation H/M/E/P	Score/ Exposure Level
6. acoustics and insulation	/ / /	/

(notations)_____
_____
_____
_____

7. frames and slabs   / / /   /

(notations)_____
_____
_____
_____

8. plenums/interstitial spaces   / / /   /

(notations)_____
_____
_____
_____

9.   / / /   /

(notations)_____
_____
_____
_____

Scoring the Evaluation

B. Exterior

Area/Item Evaluated   Risk Factor   Score/
                                    Evaluation   Exposure
(notation)   H/M/E/P   Level

1. integrity of foundations   / / /   /

(notations)_____
_____
_____
_____

2. roof system/penthouses   / / /   /

(notations)_____
_____
_____
_____

3. doors/windows/openings   / / /   /

(notations)_____
_____
_____
_____

4. appearance of facades   / / /   /

(notations)_____
_____
_____
_____

Area/Item Evaluated (notation)	Risk Factor Evaluation H/M/E/P	Score/ Exposure Level
5. adherence to building codes	/ / /	/

(notations)_____

_____

_____

_____

6. loading docks/entrances   / / /   /

(notations)_____

_____

_____

_____

7.   / / /   /

(notations)_____

_____

_____

_____

8.   / / /   /

(notations)_____

_____

_____

_____

Scoring the Evaluation

Area/Item Evaluated	Risk Factor Evaluation	Score/ Exposure Level
(notation)	H/M/E/P	

9.   / / /   /

(notations)_____
_____
_____
_____

III. SYSTEMS AND EQUIPMENT

A. General

Area/Item Evaluated	Risk Factor Evaluation	Score/ Exposure Level
(notation)	H/M/E/P	

1. preventive maintenance program   / / /   /

(notations)_____
_____
_____
_____

2. reliability testing/calibration   / / /   /

(notations)_____
_____
_____
_____

Area/Item Evaluated (notation)	Risk Factor Evaluation H/M/E/P	Score/ Exposure Level
3. ample capacity/redundancy	/ / /	/

(notations)_____
_____
_____
_____

| 4. circuit zoning for areas served | / / / | / |

(notations)_____
_____
_____
_____

| 5. labels showing directions of flow | / / / | / |

(notations)_____
_____
_____
_____

| 6. updated valve/switch lists | / / / | / |

(notations)_____
_____
_____
_____

Scoring the Evaluation

Area/Item Evaluated (notation)	Risk Factor Evaluation H/M/E/P	Score/ Exposure Level

7. identification of main shut offs   / / /   /

(notations)_____
_____
_____
_____

8. alarm system performance   / / /   /

(notations)_____
_____
_____
_____

9. duct/cable routes traced out   / / /   /

(notations)_____
_____
_____
_____

10. alternation of redundant systems / / /   /

(notations)_____
_____
_____
_____

Area/Item Evaluated (notation)	Risk Factor Evaluation H/M/E/P	Score/ Exposure Level
11. pre/post season overhauls	/ / /	/

(notations) _____
_____
_____
_____

| 12. timeliness/accuracy of repairs | / / / | / |

(notations) _____
_____
_____
_____

| 13. economic/energy efficiency | / / / | / |

(notations) _____
_____
_____
_____

| 14. time clocks and indicator lights | / / / | / |

(notations) _____
_____
_____
_____

Scoring the Evaluation

Area/Item Evaluated	Risk Factor Evaluation	Score/ Exposure
(notation)	H/M/E/P	Level

15. proper tools available/maintained / / /   /

(notations)_____
_____
_____
_____

16. adequate spare parts/consumables / / /   /

(notations)_____
_____
_____
_____

17. control systems maintained   / / /   /

(notations)_____
_____
_____
_____

18. licenses/permits current   / / /   /

(notations)_____
_____
_____
_____

Area/Item Evaluated (notation)	Risk Factor Evaluation H/M/E/P	Score/ Exposure Level
19. safety devices/interlocks	/ / /	/

(notations)_____
_____
_____
_____

| 20. sufficiency of utilities | / / / | / |

(notations)_____
_____
_____
_____

| 21. exercise of stand-by equipment | / / / | / |

(notations)_____
_____
_____
_____

| 22. system descriptions | / / / | / |

(notations)_____
_____
_____
_____

Scoring the Evaluation

Area/Item Evaluated (notation)	Risk Factor Evaluation H/M/E/P	Score/ Exposure Level

23. housekeeping in equipment rooms / / /   /

(notations)_____
_____
_____
_____

24. start-up/shut-down procedures / / /   /

(notations)_____
_____
_____
_____

25.   / / /   /

(notations)_____
_____
_____
_____

26.   / / /   /

(notations)_____
_____
_____
_____

Area/Item Evaluated (notation)	Risk Factor Evaluation H/M/E/P	Score/ Exposure Level
27.	/ / /	/

(notations)_____
_____
_____
_____

28.	/ / /	/

(notations)_____
_____
_____
_____

29.	/ / /	/

(notations)_____
_____
_____
_____

B. HVAC

Area/Item Evaluated (notation)	Risk Factor Evaluation H/M/E/P	Score/ Exposure Level
1. CW loop freeze protection	/ / /	/

(notations)_____
_____
_____
_____

Scoring the Evaluation

Area/Item Evaluated	Risk Factor Evaluation	Score/ Exposure
(notation)	H/M/E/P	Level

2. steam trap maintenance program / / /   /

(notations)_____
_____
_____
_____

3. air supply/exhaust system / / /   /

(notations)_____
_____
_____
_____

4. frequency of filter changes / / /   /

(notations)_____
_____
_____
_____

C. ELECTRICAL

Area/Item Evaluated	Risk Factor Evaluation	Score/ Exposure
(notation)	H/M/E/P	Level

1. loading/balancing of circuits / / /   /

(notations)_____
_____
_____
_____

Area/Item Evaluated (notation)	Risk Factor Evaluation H/M/E/P	Score/ Exposure Level
2. sufficient electrical outlets	/ / /	/

(notations)_____
_____
_____
_____

3. panels marked for areas served	/ / /	/

(notations)_____
_____
_____
_____

4. lighting systems maintenance	/ / /	/

(notations)_____
_____
_____
_____

5. tag/lock out program	/ / /	/

(notations)_____
_____
_____
_____

Scoring the Evaluation   145

Area/Item Evaluated	Risk Factor Evaluation	Score/ Exposure
(notation)	H/M/E/P	Level

6. emergency generator/UPS systems   / / /   /

(notations)_____
_____
_____
_____

7.   / / /   /

(notations)_____
_____
_____
_____

8.   / / /   /

(notations)_____
_____
_____
_____

9.   / / /   /

(notations)_____
_____
_____
_____

D. PLUMBING

Area/Item Evaluated (notation)	Risk Factor Evaluation H/M/E/P	Score/ Exposure Level
1. sufficient plumbing fixtures	/ / /	/

(notations)_____
_____
_____
_____

2. drains free and clear   / / /   /

(notations)_____
_____
_____
_____

3.   / / /   /

(notations)_____
_____
_____
_____

4.   / / /   /

(notations)_____
_____
_____
_____

IV. SAFETY AND FIRE PROTECTION

A. GENERAL SAFETY

Area/Item Evaluated	Risk Factor Evaluation	Score/ Exposure
(notation)	H/M/E/P	Level

1. emergency telephone list posted   / / /   /

(notations)_____
_____
_____
_____

2. safety programs reviewed annually   / / /   /

(notations)_____
_____
_____
_____

3. safety posters utilized   / / /   /

(notations)_____
_____
_____
_____

4. employee orientation/training   / / /   /

(notations)_____
_____
_____
_____

Area/Item Evaluated (notation)	Risk Factor Evaluation H/M/E/P	Score/ Exposure Level
5. safety gear readily available	/ / /	/
6. updated emergency plans	/ / /	/
7. key distribution controlled	/ / /	/
8. machine guards in place	/ / /	/

(notations) _____

(notations) _____

(notations) _____

(notations) _____

Scoring the Evaluation

Area/Item Evaluated	Risk Factor Evaluation	Score/ Exposure
(notation)	H/M/E/P	Level

9. evacuation signage / / /   /

(notations)_____
_____
_____
_____

10. disaster procedures manual / / /   /

(notations)_____
_____
_____
_____

11.   / / /   /

(notations)_____
_____
_____
_____

12.   / / /   /

(notations)_____
_____
_____
_____

Area/Item Evaluated   Risk Factor   Score/
                            Evaluation   Exposure
(notation)   H/M/E/P   Level

13.   / / /   /
(notations)_____
_____
_____
_____

14.   / / /   /
(notations)_____
_____
_____
_____

B. FIRE PROTECTION

Area/Item Evaluated   Risk Factor   Score/
                            Evaluation   Exposure
(notation)   H/M/E/P   Level

1. flammables properly stored   / / /   /
(notations)_____
_____
_____
_____

2. fire warning system/components / / /   /
(notations)_____
_____
_____
_____

Scoring the Evaluation

Area/Item Evaluated	Risk Factor Evaluation	Score/ Exposure
(notation)	H/M/E/P	Level

3. fire extinguisher inspections   / / /   /

(notations)_____

_____

_____

_____

4. flame spread ratings of finishes   / / /   /

(notations)_____

_____

_____

_____

5. fire drills conducted/critiqued   / / /   /

(notations)_____

_____

_____

_____

6. rating of door/wall assemblies   / / /   /

(notations)_____

_____

_____

_____

Area/Item Evaluated (notation)	Risk Factor Evaluation H/M/E/P	Score/ Exposure Level
7. exit access/illumination	/ / /	/

(notations)_____
_____
_____
_____
_____

| 8. slab/wall penetrations filled in | / / / | / |

(notations)_____
_____
_____
_____
_____

| 9. | / / / | / |

(notations)_____
_____
_____
_____
_____

V. ENVIRONMENTAL COMPLIANCE

A. REGULATORY

Area/Item Evaluated (notation)	Risk Factor Evaluation H/M/E/P	Score/ Exposure Level
1. stack and flue emissions monitored	/ / /	/

(notations)_____
_____
_____
_____

Scoring the Evaluation

Area/Item Evaluated	Risk Factor Evaluation	Score/ Exposure
(notation)	H/M/E/P	Level

2. hazardous materials properly stored   / / /   /

(notations)_____
_____
_____
_____

3. waste properly disposed of   / / /   /

(notations)_____
_____
_____
_____

4. potable water supplies monitored   / / /   /

(notations)_____
_____
_____
_____

5. asbestos removal/encapsulation   / / /   /

(notations)_____
_____
_____
_____

Area/Item Evaluated (notation)	Risk Factor Evaluation H/M/E/P	Score/ Exposure Level
6. effluent discharges monitored	/ / /	/

(notations)_____
_____
_____
_____

7. / / / /

(notations)_____
_____
_____
_____

8. / / / /

(notations)_____
_____
_____
_____

9. / / / /

(notations)_____
_____
_____
_____

Scoring the Evaluation

B. GENERAL

Area/Item Evaluated (notation)	Risk Factor Evaluation H/M/E/P	Score/ Exposure Level
1. noise/odors/nuisances	/ / /	/

(notations)_____
_____
_____
_____

| 2. contingency plans are in place | / / / | / |

(notations)_____
_____
_____
_____

| 3. indoor air quality | / / / | / |

(notations)_____
_____
_____
_____

| 4. | / / / | / |

(notations)_____
_____
_____
_____

IV. DEPARTMENT OPERATIONS

A. ORGANIZATION

Area/Item Evaluated (notation)	Risk Factor Evaluation H/M/E/P	Score/ Exposure Level
1. goals and objectives	/ / /	/

(notations)_____

_____

_____

_____

2. regular department meetings held / / /   /

(notations)_____

_____

_____

_____

3. technical/code libraries   / / /   /

(notations)_____

_____

_____

_____

4. blueprint/manufacturers' indexes / / /   /

(notations)_____

_____

_____

_____

Scoring the Evaluation

Area/Item Evaluated (notation)	Risk Factor Evaluation H/M/E/P	Score/ Exposure Level
5. current policies and procedures	/ / /	/

(notations)_____
_____
_____
_____

6. service agreements	/ / /	/

(notations)_____
_____
_____
_____

7. Operating Procedures Manual	/ / /	/

(notations)_____
_____
_____
_____

8. attendance at committee meetings	/ / /	/

(notations)_____
_____
_____
_____

Area/Item Evaluated (notation)	Risk Factor Evaluation H/M/E/P	Score/ Exposure Level
9. work requisition system	/ / /	/
10. project planning function	/ / /	/
11. schedules/forms	/ / /	/
12. safety/surveillance programs	/ / /	/

(notations) _____

(notations) _____

(notations) _____

(notations) _____

Scoring the Evaluation

Area/Item Evaluated (notation)	Risk Factor Evaluation H/M/E/P	Score/ Exposure Level
13. diagnostic equipment	/ / /	/

(notations)_____
_____
_____
_____

| 14. | / / / | / |

(notations)_____
_____
_____
_____

B. PERSONNEL

Area/Item Evaluated (notation)	Risk Factor Evaluation H/M/E/P	Score/ Exposure Level
1. adequacy of staffing	/ / /	/

(notations)_____
_____
_____
_____

| 2. supervisory development | / / / | / |

(notations)_____
_____
_____
_____

Area/Item Evaluated	Risk Factor Evaluation	Score/ Exposure
(notation)	H/M/E/P	Level

3. employee training   / / /   /

(notations)_____
_____
_____
_____

4. personnel files current   / / /   /

(notations)_____
_____
_____
_____

5. job description updated   / / /   /

(notations)_____
_____
_____
_____

6. written routines/procedures   / / /   /

(notations)_____
_____
_____
_____

Scoring the Evaluation

Area/Item Evaluated	Risk Factor Evaluation	Score/ Exposure
(notation)	H/M/E/P	Level

7. performance appraisals / / /   /

(notations)_____
_____
_____
_____

8. visits with off shift personnel / / /   /

(notations)_____
_____
_____
_____

9.   / / /   /

(notations)_____
_____
_____
_____

C. QUALITY OF SERVICE

Area/Item Evaluated	Risk Factor Evaluation	Score/ Exposure
(notation)	H/M/E/P	Level

1. organizational confidentiality / / /   /

(notations)_____
_____
_____
_____

Area/Item Evaluated (notation)	Risk Factor Evaluation H/M/E/P	Score/ Exposure Level
2. professional image maintained	/ / /	/

((notations)) _____
_____
_____
_____

| 3. customer relations | / / / | / |

(notations) _____
_____
_____
_____

| 4. licenses/certificates current | / / / | / |

(notations) _____
_____
_____
_____

| 5. periodic department evaluations | / / / | / |

(notations) _____
_____
_____
_____

Scoring the Evaluation

Area/Item Evaluated	Risk Factor Evaluation	Score/Exposure
(notation)	H/M/E/P	Level

6. frequent conduction of inspections / / / /

(notations)_____
_____
_____
_____

7. elimination of work back-logs / / / /

(notations)_____
_____
_____
_____
_____

8. jurisdictional compliance / / / /

(notations)_____
_____
_____
_____

9. setting goals and objectives / / / /

(notations)_____
_____
_____
_____

Area/Item Evaluated (notation)	Risk Factor Evaluation H/M/E/P	Score/ Exposure Level
10. encouragement of feedback	/ / /	/

(notations)_____
_____
_____
_____

11. safety rules enforced   / / /   /

(notations)_____
_____
_____
_____

12. housekeeping required   / / /   /

(notations)_____
_____
_____
_____

13.   / / /   /

(notations)_____
_____
_____
_____

Scoring the Evaluation

Area/Item Evaluated	Risk Factor	Score/
	Evaluation	Exposure
(notation)	H/M/E/P	Level

14.   / / /   /

(notations)_____
_____
_____
_____

D. RETENTION OF RECORDS

Area/Item Evaluated	Risk Factor	Score/
	Evaluation	Exposure
(notation)	H/M/E/P	Level

1. employee training/orientation   / / /   /

(notations)_____
_____
_____
_____

2. equipment operating logs   / / /   /

(notations)_____
_____
_____
_____

Area/Item Evaluated (notation)	Risk Factor Evaluation H/M/E/P	Score/ Exposure Level
3. equipment repair histories	/ / /	/

(notations)_____
_____
_____
_____

4. utility consumption	/ / /	/

(notations)_____
_____
_____
_____

5. personnel jackets	/ / /	/

(notations)_____
_____
_____
_____

6. work/vacation schedules	/ / /	/

(notations)_____
_____
_____
_____

Scoring the Evaluation

| Area/Item Evaluated (notation) | Risk Factor Evaluation H/M/E/P | Score/ Exposure Level |

7. inspection/compliance reports   /   /   /   /

(notations)_____
_____
_____
_____

8. valve/switch lists   /   /   /   /

(notations)_____
_____
_____
_____

9. capital/operating budget variances /   /   /   /

(notations)_____
_____
_____
_____

10. updated job descriptions   /   /   /   /

(notations)_____
_____
_____
_____

Area/Item Evaluated	Risk Factor	Score/
	Evaluation	Exposure
(notation)	H/M/E/P	Level

11. meeting minutes   / / /   /

(notations)_____
_____
_____
_____

12. operating certificates   / / /   /

(notations)_____
_____
_____
_____

13. incident reports/investigations   / / /   /

((notations)_____
_____
_____
_____

14. machinery histories   / / /   /

(notations)_____
_____
_____
_____

Scoring the Evaluation

Area/Item Evaluated	Risk Factor Evaluation	Score/ Exposure
(notation)	H/M/E/P	Level

15. temperature/humidity logs   / / /   /

(notations)_____
_____
_____
_____

16. contract services performed   / / /   /

(notations)_____
_____
_____
_____

17. supply/equipment inventories   / / /   /

(notations)_____
_____
_____
_____

18.   / / /   /

(notations)_____
_____
_____
_____

Area/Item Evaluated (notation)	Risk Factor Evaluation H/M/E/P	Score/ Exposure Level
19.	/ / /	/

(notations)_____
_____
_____
_____

20.   / / /   /

(notations)_____
_____
_____
_____

21.   / / /   /

(notations)_____
_____
_____
_____

22.   / / /   /

(notations)_____
_____
_____
_____

Scoring the Evaluation   171

Area/Item Evaluated   Risk Factor   Score/
                               Evaluation   Exposure
(notation)   H/M/E/P   Level

23.   / / /   /

(notations)_____
_____
_____
_____

24.   / / /   /

(notations)_____
_____
_____
_____

E. OTHER

Area/Item Evaluated   Risk Factor   Score/
                               Evaluation   Exposure
(notation)   H/M/E/P   Level

1.   / / /   /

(notations)_____
_____
_____
_____

Area/Item Evaluated (notation)	Risk Factor Evaluation H/M/E/P	Score/ Exposure Level
2.	/ / /	/

(notations)_____
_____
_____
_____

3.　　　　　　　　　　　　　　／ ／ ／　　　／

(notations)_____
_____
_____
_____

4.　　　　　　　　　　　　　　／ ／ ／　　　／

(notations)_____
_____
_____
_____

5.　　　　　　　　　　　　　　／ ／ ／　　　／

(notations)_____
_____
_____
_____

Scoring the Evaluation

Area/Item Evaluated	Risk Factor Evaluation	Score/ Exposure
(notation)	H/M/E/P	Level

6.   / / /   /

(notations)_____
_____
_____
_____
_____

7.   / / /   /

(notations)_____
_____
_____
_____
_____

8.   / / /   /

(notations)_____
_____
_____
_____
_____

9.   / / /   /

(notations)_____
_____
_____
_____
_____

Area/Item Evaluated (notation)	Risk Factor Evaluation H/M/E/P	Score/ Exposure Level
10. (notations)_____	/ / /	/
11. (notations)_____	/ / /	/
12. (notations)_____	/ / /	/
13. (notations)_____	/ / /	/
14. (notations)_____	/ / /	/

CHAPTER 10
THE EVALUATION REPORT AND CORRECTIVE ACTION PLAN

Hold on a second... I warned you the evaluation process was long and involved, and certainly I empathize with you but it won't do to get PO'ed over your POE. Nor, as you're wont to report, is it quite finished. There still remains the business of correlating and analyzing the data, transforming it into a readable document and deciding how you'll follow up on your findings. Shall we begin... ?

LAYING OUT THE REPORT

The instrument you design for conveying the evaluation results depends on several factors, high among them, your bent for writing and the intended reader. My feeling is (and experience bears this out), that the simpler you make it, the more likely people will read and heed its messages. Figures 10-1a thru c show how my personally created format has evolved over time. Although the intensity and sophistication of the evaluations performed has increased, aside from some minor modifications, the reporting mechanism hasn't changed that much, the idea being to make it easy for the people holding the purse strings to follow.

INTRODUCING AND
SUMMARIZING THE EVALUATION

As with the remainder of the report, its introduction and summarization should be made as short as possible. If you want the introduction

Figure 10-1a. Evolution of an evaluation report

TABLE OF CONTENTS

Page

I. EXTERIOR BUILDING & GROUNDS

    A. Existing Conditions . . . .
        1. parking lots . . . .
        2. ways of travel . . . .

    B. Utilities/equipment . . . .
    C. Outbuildings/off site . . . .

II. PHYSICAL PLANT

    A. Interior structures . . . .
        1. existing conditions . . . .
        2. life safety code . . . .
        3. support system priorities . . . .

    B. Equipment/system overview . . . .
        1. preventive maintenance . . . .
        2. documentation . . . .
        3. spares/parts inventory . . . .
        4. nuisance problems . . . .

III. DEPARTMENT MANAGEMENT

    A. Department Objectives . . . .

    B. Personnel
        1. staffing . . . .
        2. competency/training . . . .
        3. employee comments . . . .

    C. Operations
        1. corrective maintenance . . . .
        2. rounds/routines . . . .
        3. materials acquisition . . . .
        4. policies/procedures . . . .
        5. service agreements . . . .

    D. Work Backlog . . . .

IV. REGULATORY ISSUES

V. AUTHORS OPINION

VI. RECOMMENDATIONS SUMMARY

Figure 10-1b. Evolution of an evaluation report

```
                        TABLE OF CONTENTS


    I.   EXTERIOR BUILDING/GROUNDS

         A.  Existing Conditions

         B.  Outdoor Utilities/Equipment

         C.  Outbuilding/Satellites


    II.  PHYSICAL PLANT

         A.  Interior Structures

         B.  Equipment/System Overview

         C.  Preventive Maintenance


    III. PLANT OPERATIONS DEPARTMENT

         A.  Department Objectives

         B.  Organization

         C.  Operations


    IV.  REGULATORY/ENVIRONMENTAL ISSUES

         A.  Asbestos Abatement

         B.  PCB Contaminated Transformers

         C.  Underground Oil tanks

         D.  Incinerator Air Emissions

         E.  Other Issues


    V.   SUMMARY
```

Figure 10-1b. Evolution of an evaluation report (cont'd)

VI. SUPPORTING DOCUMENTATION

    APPENDIX

    A. Roof Survey

    B. Underground Oil Tanks

    C. Light Standard Survey

    D. Power House Floor Survey

    E. Wall Penetration Report

    F. Floor Tile Replacement

    G. Door Survey

    H. Generator Assessment Report

    I. Incinerator Repair Report

    J. Elevator Full Load Test Report

    K. Dish Machine Repair Estimate

    L. Electrical Distribution System Survey

    M. Frozen Coil History

    N. Preventive Maintenance Plan

    O. Department Reorganization Inquiry

    P. Request for Staffing Increase

    Q. Department Computerization Justification

    R. Work Order System Description

    S. S.O.P. Manual Outline

    T. Reclassification of Personnel

VII. RECOMMENDATIONS/ACTION PLAN

Figure 10-1c. Evolution of an evaluation report

```
                        CONTENTS


1.      disclaimer


I.      SITE CHARACTERISTICS

        A.  Existing conditions
        B.  Grounds maintenance

II.     BUILDING STRUCTURES

        A.  Interior
        B.  Exterior

III.    SYSTEMS AND EQUIPMENT

        A.  General
        B.  HVAC
        C.  Electrical
        D.  Plumbing

IV.     SAFETY AND FIRE PROTECTION

        A.  General safety
        B.  Fire protection

V.      ENVIRONMENTAL COMPLIANCE

        A.  Regulatory
        B.  Internal

VI.     DEPARTMENT OPERATIONS

        A.  Organizational
        B.  Personnel
        C.  Quality of service
        D.  Records retention

VII.    EVALUATION SUMMARY

VIII.   SUPPORTING DOCUMENTATION

IX.     CORRECTIVE ACTION PLAN
```

read, it shouldn't be more than a paragraph long. But it can be concise and still carry a big message. The introduction I often use is in the form of this disclaimer:

INTRODUCTION

This assessment was prompted by the obvious need for restructuring the Engineering/Maintenance Department. It is intended to serve as an objective, cursory diagnosis of deficiencies existing within the physical plant, as a preliminary evaluation of the company's maintenance operations and to function as a data base of problematical information from which corrective plans of action can be formulated. The recommendations contained herein were made after reviewing scant, quickly derived and sometimes questionable data which may later prove unreliable and must not be construed as the "last word" on any of the issues addressed.

Kenneth Lee Petrocelly
RPA/SMA/CPE/CHE/FMA
Facilities Manager

TRANSCRIBING THE SCORECARD

The actual composition of the evaluation report simply entails consulting the scorecard, reviewing support documents, recalling observations made and arranging the data in a predetermined format. The outline I prefer is to cite a general observation; list several specific, witnessed instances relating to it; draw a conclusion; then make recommendations for correcting the problems. For example, under section I. Site Characteristics, B. Ground Maintenance...

Observation #1
The aesthetic appearance of the grounds project a less than desirable image to the general public.

specific
- there are several dead trees on the property that need cut down and removed

- bushes need trimmed/trees need pruning
- the grounds are cluttered with conduits and building materials in several locations
- debris from contractor work is evident
- paper wrappers/cups... etc. scattered through grassy areas
- road salt is saturating the dirt in the grassy areas
- the maintenance section has no groundskeeping expertise on board
- the maintenance section is not equipped for groundskeeping activity
- no sprinkler system in place for watering
- weeds growing out of cracks in concrete
- leaves accumulated in base of stairwells
- grass extending over walkways need edged
- mulching around bushes/trees disturbed or thinning
- barrels and trash cluttering fenced-in compactor area
- shabby walk off mats at entrances need replaced
- no provisions for attractive salt bins at entrances
- branches from trees strewn around property
- no groundskeeping policies presently exist

conclusion
A beautification program should be established to restore the grounds to their original countenance

RECOMMENDATION NO. 1

Hire one full-time groundskeeper who has the knowledge and experience necessary to establish and carry out a comprehensive grounds care program

RECOMMENDATION NO. 2

Purchase and provide storage space for the items necessary to institute a groundskeeping program

RECOMMENDATION NO. 3

Write policies/procedures addressing routines, rounds, winterization programs, snow removal, lawn, shrub care... etc.

... followed by a second observation, third and so on until the category has been exhausted.

CORRELATING THE SUPPORT DATA

In Figure 10-1b, the Table of Contents indicates that all the documentation supporting the body of the report can be found in Section VI. As you can infer from the titles listed there, the appendix is a hodgepodge of data, fixed in a central location, which is meant to be referred to in support of and while pursuing the evaluation report. Any information which clarifies or lends credibility to the report can be stored there, including...

- letters of commitment
- quotations and estimates
- surveys and studies
- sketches or blueprints
- copies of invoices
- excerpts from applicable laws
- inspection reports
- various correspondence
- policies and procedures
- operating instructions
- notices and questionnaires
- chronologies and histories
- other supporting documentation

FASHIONING THE CORRECTIVE ACTION PLAN

Once all the appropriate people have been apprised of the organization's physical, functional and operational deficiencies, the powers that be must then get together and develop a plan of action to correct the maladies. At this point, they are cognizant of what discrepancies exist, the extent of their severity and, if you've explained it adequately, the consequences of inattention to the problems, on their part. All that's left hanging is their prioritization and resolution.

Some of the recommendations need not even be included in the corrective action plan (i.e., items such as replacement of burned-out indicator lamps or those whose expense will be insignificant enough that they can be included in your department budget), or they may be relegated to a special section of the instrument under the heading of "maintenance items." The remainder of them should be listed either by way of descending level of importance as in Figure 10-2a or in accord with pre-established levels as in Figure 10-2b (see pages 184-185).

At a minimum, the corrective action plan should include the recommendations extracted from the evaluation report in some order of priority, a listing of the resources needed for compliance with each recommendation, an estimated cost in dollars and manpower, and an indication of how long it will take for its completion.

Well, as they say, that about wraps things up. Good luck with the Building/Appropriations Committee... and, oh yeah! Don't forget your Maalox.

Figure 10-2a. Corrective Action Plan Examples

PLAN OF ACTION

(compliance schedule for recommendations taken from the Physical Plant Operations August 1988 department assessment)

PRIORITY	REC #	RECOMMENDATION	RESOURCES NEEDED	EST. COMPLETION
1	47	Develop a policy for use of Physical Therapy Department equipment by non-patient personnel	Plant Operations/Physical Therapy Depts	Sept 88
2	48	Develop policies that address access to the emergency area by vehicles	Plant Operations Dept	Sept 88
3	49	Have Departments safety manual reviewed by the Safety Committee	Plant Operations Secy Safety Committee	Sept 88
4	50	Implement a hazard surveillance program which reports findings to the Safety Committee	Plant Operations Dept Risk Manager Safety Committee	Sept 88
5	51	Monitor the results of safety programs for their effectiveness at least annually	Plant Operations Dept Safety Committee	Ongoing
6	52	Contingency plans need to be developed in the event of the loss of communications systems in the hospital - this includes telephone, nurse call and paging systems	Plant Operations Dept	Sept 88
7	53	Fire Drill critiques do not show corrective actions taken	Safety Committee	—
8	54	All individuals are in-serviced on the use and handling of hazardous materials needs documented	Plant Operations Dept Risk Manager	Ongoing
9	55	Include the nurse call and paging systems in the communication system PM program	Plant Operations Dept Computerized PM program	Sept 88
10	56	Develop alternate means of communications for systems where failure can have life threatening consequences	Plant Operations Dept	Sept 88
11	57	Install a call bell in the laboratory patient restroom	Maintenance Section	Sept 88

Figure 10-2b. Corrective Action Plan Examples

FACILITIES EVALUATION ACTION PLAN

DATE 1/2/90
BY

#	RECOMMENDATION	*	RESOURCES REQUIRED	ESTIMATED COST	COMPLETION TARGET
1	HIRE 1 FULL TIME GROUNDSKEEPER WHO HAS THE KNOWLEDGE AND EXPERIENCE NECESSARY TO ESTABLISH AND CARRY OUT A COMPREHENSIVE GROUNDS CARE PROGRAM	C	Administrative Approval Human Resources Dept. Plant Operations Dept -Maint. Section	$23,600.00 base wages plus benefits per annum	
2	PURCHASE AND PROVIDE STORAGE SPACE FOR THE ITEMS NECESSARY TO INSTITUTE A GROUNDSKEEPING PROGRAM	C	Administrative Approval Facilities Manager	$10,000.00	
3	WRITE POLICIES/PROCEDURES ADDRESSING ROUTINES, ROUNDS, WINTERIZATION PROGRAMS, SNOW REMOVAL, LAWN/SHRUB CARE....etc.	B	Administrative Approval Plant Operations Dept. -Maintenance Section -Clerical	—	
4	HIRE OUT A VENDOR TO FACILITATE A FIX UP/PAINT UP/CLEAN UP OF THE BUILDING PROPER	C	Administrative Approval Facilities Manager Outside Contractors	$100,000.000 ($35,000.00 for replacement of light standards)	

*(Priority____   A=Urgent   B=Necessary   C=Desirable   D=Deferred)

Appendix A
Safety Committee Charter

After due scrutiny and discussion, the members of the Safety Committee, by unanimous consent, hereby adopt this charter to serve as its means for establishing and insuring the safety integrity of the Hospital.

The undersigned recognize the charter as the official operating standard of the Safety Committee and approve it as written.

_____ _____
Chairman, Safety Committee   Date
Safety Director

_____ _____
Administrator   Date

_____ _____
Chief of Medical Staff   Date

SAFETY COMMITTEE CHARTER

PURPOSE:
To document the principles, rights, aims and duties of the Safety Committee.

PHILOSOPHY:
It is the intent of the Safety Committee to establish and maintain the highest level of environmental safety obtainable, affecting all persons, within the confines or on the premises of Hospital.

AUTHORITY:
The Safety Committee, through its Chairperson, is empowered under the auspices of the Administrator and the Medical Staff, to institute control measures or studies made necessary when a hazardous condition exists that, in their judgement, could result in personal injury to people or damage to hospital equipment, buildings, property or systems.

STRUCTURE:
The Chairperson of the Committee shall be the Safety Director of the Hospital. A Co-Chairperson will be designated as a stand-in for the Chairperson in his/her absence. The Committee shall be comprised of, but not limited to representatives from Administration, Medical Staff, Nursing, Plant Operations, Environmental Services, Dietary Services, Employee Health/Infection Control, Human Resources and Quality Assurance. All members of the Committee will be appointed by the Administrator with the exception of the Medical Staff who will be appointed by the Chief of the Medical Staff.

A secretary shall preside at all meetings for the purpose of recording minutes.

OBJECTIVES:
1. Adherence to and implementation of the Standards for Plant Technology and Safety Management as outlined in the Accreditation Manual for Hospitals issued by the Joint Commission on Accreditation for Hospitals as follows:

Appendix A — Committee Charter

STANDARD I - Buildings in which patients are housed overnight or receive treatment are designed, constructed, equipped and maintained in a manner that is designed to provide for the physical safety of patients, personnel, and visitors.

STANDARD II - Facility grounds are maintained in a manner that is designed to provide safe access to and a safe environment for patients, personnel, and visitors.

STANDARD III - The hospital has a system for providing a safe environment for patients, personnel, and visitors and for monitoring that environment.

STANDARD IV - The hospital has an organized safety education program.

STANDARD V - The hospital has an emergency preparedness program designed to provide for the effective utilization of available resources so that patient care can be continued during a disaster.

STANDARD VI - There is a system that is designed to safety manage hazardous materials and waste.

STANDARD VII - The hospital is provided with fire safety and other safety systems that are designed, constructed, installed, and maintained in a manner that is designed to protect patients, personnel, visitors, and property from fire and the products of combustion.

STANDARD VIII - There are communication systems designed to operate effectively and reliably. The systems are designed for the management of routine and emergency operations consistent with the services provided by the hospital.

STANDARD IX - Where appropriate, the hospital has a program designed to assure that patient care equipment, whether electrically or non-electrically powered, performs properly and safely and that individuals are trained to operate the equipment they use in performance of prescribed duties.

STANDARD X - The hospital has a program designed to assure that non-patient care, electrically powered, line-operated equipment is electrically safe.

STANDARD XI - The hospital has an electrical distribution system that is designed, installed, operated, and maintained in a manner that is designed to provide electrical power for all required operations.

STANDARD XII - The hospital has a reliable, adequately sized emergency power system.

STANDARD XIII - Where provided, vertical and horizontal transport systems are designed, installed, operated, and maintained in a manner that is designed to provide for safe transport.

STANDARD XIV - Where provided, the heating, ventilating and air conditioning (HVAC) system is designed, installed, operated, and maintained in a manner that is designed to provide a comfortable and safe environment for patients, personnel, and visitors.

STANDARD XV - The plumbing systems are designed, installed, operated, and maintained in a manner that is designed to provide an adequate and safe supply of water for all required facility operations and to facilitate the complete and safe removal of all storm water and waste water.

Appendix A — Committee Charter

STANDARD XVI - Where provided, boiler systems are designed, installed, operated, and maintained in a manner that is designed to provide an adequate and safe supply of steam and/or hot water for all required facility operations.

STANDARD XVII - Where provided, medical-gas systems are designed, installed, operated, and maintained in a manner that is designated to provide an adequate and safe supply of nonflammable medical gases for all required hospital operations.

STANDARD XVIII - Where provided, medical/surgical vacuum systems are designed, installed, operated and maintained in a manner that is designed to provide an adequate and safe level of suction for all required medical and surgical needs.

STANDARD XIX - There are safety devices and operational practices for the safety of patients and personnel.

2. Adherence to and Implementation of all safety standards as outlined in the Accreditation Requirements of the American Osteopathic Association as follows:

III. HOSPITAL PHYSICAL PLANT

    A. REQUIREMENTS AND INTERPRETATION
The hospital shall comply with all State and local laws and regulations relating to patient care and safety, sanitation and fire protection. Diagnostic and therapeutic facilities, supplies and equipment shall assure the highest level of patient care to be provided by professional and nursing staffs.
       1. Plant Construction
          a. The hospital plant shall have current approval by appropriate state and/or local authorities.

b. The physical plant shall be kept clean and in good repair.
c. All patients' rooms shall conform to acceptable hospital standards.
d. A hospital wishing to become accredited shall meet the safety standards of the Life Safety Code currently accepted by the Department of Health, Education and Welfare.
2a. To embrace all other department-specific safety requirements as listed in the latest edition of the Accreditation Requirements Manual of the American Osteopathic Association.
3. Strict enforcement of the National Fire Codes.
4. To work within generally accepted engineering practices of applicable local, state and federal agencies, relevant to safety procedures and policies in the building and health care fields.

GENERAL RESPONSIBILITIES:

It shall be the responsibility of the Safety Committee, through its Chairperson to:

1. maintain a liaison with the Medical Staff, Radiology, Infection Control, Disaster and Standards Committee and any other group or persons deemed necessary by them to insure that its objectives are met.
2. familiarize themselves with local, state and federal safety regulations applicable to the hospital.
3. maintain a library of state-of-the-art publications on hospital safety.
4. maintain written documentation of the safety violations or potential hazards found in the hospital, which includes appropriate comments and/or recommendations for correcting the problems.
5. encourage safety education programs and monitor all safety-related training activities that are established for and by

hospital personnel, including information used in new employee orientation and continuing education.
6. attend quarterly safety meetings wherein:
   a. membership attendance is mandatory unless an acceptable excuse is presented.
   b. one-third of the membership shall constitute a quorum.
   c. minutes shall be kept at each meeting which will be typed and distributed to the Administrator, Medical Staff secretary and committee members for review and comment prior to the next scheduled meeting.
7. recommend general hospital safety policies.
8. annually review specific departmental safety policies.
9. annually review the Disaster and Fire Safety programs.
10. semi-annually evaluate itself to determine its effectiveness as a safety body.
11. review all significant safety incidents involving patients, visitors and employees.
12. quarterly review safety incident summaries and determine what action must be taken to improve on the rate and severity of occurrences.
13. investigate and take corrective action on serious accidents and fires; then report findings and action taken in writing, to Administration and the department involved, as appropriate.
14. make monthly safety inspections of the facility and:
    a. take immediate action to correct discrepancies in emergency situations.
    b. review the findings at the next regularly scheduled safety meeting wherein the committee will establish a time frame for correction of the problem.
    c. submit a written memo to the Department Head indicating the nature of the problem, the date by which it must be corrected and any recommendations made by the committee which may help to resolve the matter.
    d. check to ascertain that the problem has been resolved and duly noted in the safety manual.

SPECIFIC DUTIES OF THE CHAIRPERSON:

It shall be the responsibility of the chairperson to:

1. conduct safety meetings as necessary but not less than quarterly.
2. report pertinent committee findings and recommendations in writing to Administration and the Medical Staff.
3. act as the safety liaison to the Administrator.
4. designate a co-chairperson to act in the stead of the chairperson during his/her absence.
5. assign committee members specific duties.
6. oversee sub-committee functions.
7. assure the safety inspections are performed monthly.
8. oversee the safety investigation of accidents and incidents, and determine for presentation to the committee, possible corrective measures.
9. assure compliance of all established programs.
10. coordinate safety activities with other committees.
11. act as a staff resource to departments, Administration and Board of Directors with respect to all safety matters; acting as the hospital's interface with safety-oriented agencies.

SPECIFIC DUTIES OF THE MEMBERS

It shall be the responsibility of the members to:

1. attend all scheduled meetings or notify the committee secretary of their anticipated absence.
2. serve on sub-committees and study groups as required by decision of the main body.
3. accept assignments from the Chairperson concerning committee business.
4. research and provide informational input to the main body for use in conclusion drawing and decision making.

SPECIFIC DUTIES OF THE SECRETARY

It shall be the responsibility of the Secretary to:
1. notify all members of scheduled meetings one week prior to their occurrence.
2. type and distribute meeting agenda.
3. attend and record minutes of all safety meetings.
4. maintain attendance records.
5. type and distribute copies of the minutes to the members, Medical Staff and Administrative secretaries within one week following the meeting.

Appendix B
Supervisors' Safety Hazards Worksheets

ELECTRICAL HAZARDS

Building: _____

Area: _____

OBSERVATION	YES	NO
All tools and appliances are either double insulated or electrically grounded.	[ ]	[ ]
Circuits are de-energized, locked out and tagged prior to performing electrical repairs.	[ ]	[ ]
Extension cords are properly sized and only used on a temporary basis for pre-approved work.	[ ]	[ ]
Defective electrical devices are immediately taken out of service, reported and repaired or replaced.	[ ]	[ ]
Miscellaneous electrical apparatus are well supported, and kept free of grease and dirt.	[ ]	[ ]
Only explosion proof appliances and fittings are used in combustible atmospheres.	[ ]	[ ]
Electrical rooms are uncluttered, dry and all switch panels are enclosed or guarded.	[ ]	[ ]

Findings: _____
_____
_____

Inspected by: _____ Date: _____
Reported to: _____ Date: _____
Action taken: _____
_____
_____

Appendix B — Supervisors' Safety Hazards Worksheets

EQUIPMENT HAZARDS

Building: _____

Area: _____

OBSERVATION	YES	NO
Machine guards are in place and functional.	[ ]	[ ]
Valves and switches operate easily and are labeled for the areas they serve.	[ ]	[ ]
Compressed gas cylinders are properly labeled and secured out of the way in an upright position.	[ ]	[ ]
Start up, operating and shut down procedures are posted prominently by the equipment they describe.	[ ]	[ ]
All equipment is in good working order.	[ ]	[ ]
Contingency lighting systems are frequently checked to assure their operation during emergencies.	[ ]	[ ]
Equipment and vehicle fuel supplies are maintained under lock during off-duty hours.	[ ]	[ ]

Findings: _____
_____
_____

Inspected by: _____ Date: _____
Reported to: _____ Date: _____
Action taken: _____
_____
_____

FIRE HAZARDS

Building: _____

Area: _____

OBSERVATION   YES   NO

Flammable materials are stored in safety   [ ]   [ ]
containers and are properly labeled.

Fire extinguishers are adequate in type and size for   [ ]   [ ]
the areas served, properly hung and easily accessible.

Areas are kept free on oily and or solvent   [ ]   [ ]
soaked rags and combustible waste products.

Safe techniques are practiced when transport-   [ ]   [ ]
ing, transferring or handling combustibles.

Leakage of combustible gases and/or   [ ]   [ ]
liquids is immediately reported and corrected.

Fire fighting equipment and system components   [ ]   [ ]
are unobstructed and carry current inspection tags.

Adequate ventilation is provided in areas wherein   [ ]   [ ]
flammable mists/vapors are generated or any form
of combustion occurs.

Findings: _____
_____
_____
_____

Inspected by: _____ Date: _____
Reported to: _____ Date: _____
Action taken: _____
_____
_____

Appendix B — Supervisors' Safety Hazards Worksheets

PROGRAM COMPLIANCE

Building: _____

Area: _____

OBSERVATION	YES	NO
Weekly safety meetings are held with department employees.	[ ]	[ ]
Employees' attendance at company safety in-services is current.	[ ]	[ ]
Fire and disaster drills have been conducted, critiqued and recorded.	[ ]	[ ]
Accidents/incidents are immediately reported, investigated and followed up.	[ ]	[ ]
The company smoking policy is strictly adhered to.	[ ]	[ ]
All safety equipment meets existing safety requirements and is regularly tested.	[ ]	[ ]

Findings: _____
_____
_____

Inspected by: _____ Date: _____
Reported to: _____ Date: _____
Action taken: _____
_____
_____

SECURITY

Building: _____

Area: _____

OBSERVATION	YES	NO
Adequate security measures are in place and uniformly applied.	[ ]	[ ]
Chronological records are kept of all plant visitations.	[ ]	[ ]
Fencing and other pedestrian barriers are maintained in good condition.	[ ]	[ ]
Only authorized openings are accessed for entering and exiting company areas	[ ]	[ ]
Alarm systems, security vehicles and equipment are operable and frequently tested.	[ ]	[ ]
Security personnel are properly trained and cognizant of the limitations of their authority.	[ ]	[ ]
The key control policy is strictly adhered to.	[ ]	[ ]
Thefts and other security infractions are immediately reported to the appropriate authorities.	[ ]	[ ]

Findings: _____
_____
_____

Inspected by: _____ Date: _____
Reported to: _____ Date: _____
Action taken: _____
_____
_____

Appendix B — Supervisors' Safety Hazards Worksheets

STRUCTURAL HAZARDS

Building: _____

Area: _____

OBSERVATION	YES	NO
Penetrations made through walls, floors and ceilings are properly closed.	[ ]	[ ]
Stairwells are well lighted, and handrails are in good repair.	[ ]	[ ]
Loading of walls, floors and ceilings are maintained within acceptable limits.	[ ]	[ ]
Shelving and bins holding supplies, materials and/or records are structurally sound and never overloaded.	[ ]	[ ]
Pallets, skids and drums are maintained in good repair, stacked neatly and never overloaded.	[ ]	[ ]
Masonry building facades are maintained structurally sound.	[ ]	[ ]

Findings: _____
_____
_____
_____

Inspected by: _____ Date: _____
Reported to: _____ Date: _____
Action taken: _____
_____
_____
_____

WORK SPACES

Building: _____

Area: _____

<u>OBSERVATION</u>   <u>YES</u>   <u>NO</u>

Lighting is adequate in type and intensity   [ ]   [ ]
for the work being performed.

Work areas are kept clean, orderly and   [ ]   [ ]
well ventilated.

Noise, temperature and humidity levels are   [ ]   [ ]
maintained within tolerable limits.

Protective clothing appropriate to work   [ ]   [ ]
being performed is worn in accordance
with company policy.

Aisles are of adequate width, well   [ ]   [ ]
marked and kept free of obstructions.

Guardrails are secured in place and   [ ]   [ ]
highly visible.

Tools are kept clean, in good repair   [ ]   [ ]
and properly stored.

Findings: _____
_____
_____

Inspected by: _____ Date: _____
Reported to: _____ Date: _____
Action taken: _____
_____
_____

Appendix C
Statement of Construction and Fire Protection

INSTRUCTIONS

1. Contents of this document shall be verified by one of the following individuals: a fire protection engineer, a certified safety professional, a qualified representative of a fire insurance rating organization, a registered professional engineer, a registered architect, the state fire marshall or equivalent individual.

2. The types of building construction considered in this document are based on the NFPA 220, Types of Building Construction, 1985.

3. Complete a Statement of Construction and Fire Protection for every building and for all major additions to these buildings which have different structural fire protection characteristics.

4. The facility shall determine which sections of the Life Safety Code (NFPA 101, 1988 Edition) are applicable.

5. A rough sketch must be completed for each Statement of Construction.

6. All questions must be answered. Indicate "Y" for yes, "N" for no, "NA" where not applicable, unless otherwise specified.

7. List on the next page all "equivalencies" granted by the authority having jurisdiction and append supporting documents.

_____ STATEMENT VERIFIED BY _____

Name	Title	Profession
Organization	Registration No.	State License No.
Street, City, State, Zip Code	(Area Code) Telephone No.	
Signature		Date

• Training, Education and Experience _____
_____
_____
_____
_____
_____
_____
_____
_____

Equivalencies _____
_____
_____
_____
_____
_____
_____
_____

AppendixC — Statement of Construction and Fire Protection

1. Give a rough sketch below showing the original and subsequent construction. Indicate approved exits with an X and 2-hour fire separations.

PLAN VIEW (Top View)

ELEVATION (Side View)

2. Was the statement completed:
   a. From available records only   Y   N
   b. After an on-site survey only   Y   N
   c. From both records and on-site survey   Y   N

3. Specific reference is made to:
   NFPA 101 (Life Safety Code) 1988
   NFPA 220, 1985

4. For this specific building or addition, state the year of final design approval by the Authority having jurisdiction. (If unknown, state year of construction.) _____

   _____

5. Indicate:
   a. The number of stories above ground in this regardless of their use (include ground level) _____

   b. The floors above ground that are occupied by the facility

      (Stories shall be counted starting at the lowest floor of exit discharge which is level with or above finished grade of the exterior wall line for 50 percent or more of its perimeter _____

6. Indicate the number of levels below ground _____

7. The basic structural characteristic of this building or addition is at least
   a. Type I 443 Fire Resistant   Y   N
   b. Type I 332 Fire Resistant   Y   N
   c. Type II 222 Fire Resistant   Y   N
   d. Type II 111 Protected Noncombustible   Y   N
   e. Type II 000 Unprotected Noncombustible   Y   N
   f. Type III 211 Protected Ordinary   Y   N
   g. Type III 200 Unprotected Ordinary   Y   N
   h. Type IV 2HH Protected Ordinary   Y   N
   i. Type V 111 Protected Wood Frame   Y   N
   j. Type V 000 Unprotected Wood Frame   Y   N
   k. Other (describe) _____

8. There is an approved automatic fire extinguishing system:
   a. Throughout the building   Y   N

Appendix C — Statement of Construction and Fire Protection

    b. Other (state location) _____

    _____

9. The principal supporting members (such as columns, beams, girders and trusses) or exterior bearing walls for one or more than one floor and/or roof, are of:
   a. At least 4-hour fire resistive construction   Y   N
   b. At least 3-hour fire resistive construction   Y   N
   c. At least 2-hour fire resistive construction   Y   N
   d. At least 1-hour fire resistive construction   Y   N
   e. Less than 1-hour fire resistive construction   Y   N
      (noncombustible materials)
   f. Less than 1-hour fire resistive construction   Y   N
      (limited combustible materials)
   g. Other _____   Y   N

10. The principal supporting members (such as columns, beams, girders and trusses) for the structural frame for one or more floors are of:
    a. At least 3-hour fire resistive construction   Y   N
    b. At least 2-hour fire resistive construction   Y   N
    c. At least 1-hour fire resistive construction   Y   N
    d. Less than 1-hour fire resistive construction   Y   N
       (noncombustible materials)
    e. Less than 1-hour fire resistive construction   Y   N
       (limited combustible materials)
    f. Other _____   Y   N

11. The floor/ceiling assemblies (secondary floor supporting members such as beams, slabs and joints) are of:
    a. At least 3-hour fire resistive construction   Y   N
    b. At least 2-hour fire resistive construction   Y   N
    c. At least 1-hour fire resistive construction   Y   N
    d. Less than 1-hour fire resistive construction   Y   N
       (noncombustible materials)

e. Less than 1-hour fire resistive construction   Y   N
   (limited combustible materials)
f. Other _____   Y   N

12. The principal means of fire protection for the supporting members is

		Vertical	Horizontal
a.	Reinforced concrete or concrete enclosed steel	[ ]	[ ]
b.	Sprayed-in fire proofing	[ ]	[ ]
c.	Metal lath and plaster or gypsum board	[ ]	[ ]
d.	Listed membrane fire proofing (for horizontal members)	[ ]	[ ]

    e. Other (describe) _____
    _____
    _____

13. Both principal and secondary roof supporting members are of:
    a. At least 2-hour fire resistive construction   Y   N
    b. At least 1-1/2-hour fire resistive construction   Y   N
    c. At least 1-hour fire resistive construction   Y   N
    d. Less than 1-hour fire resistive construction   Y   N
       (noncombustible materials)
    e. Less than 1-hour fire resistive construction   Y   N
       (limited combustible materials)
    f. Other _____
    _____

(Note: If combustible materials are used for principal or secondary roof supporting members, complete the following. If not, proceed to question number 15.)

Appendix C — Statement of Construction and Fire Protection

g. The roof coverings meet Class C requirements Y N
h. The roof is separated from the rest of the building by a noncombustible floor assembly which includes at least 2-1/2 inches of concrete or gypsum. Y N
i. All penetrations of the separation between the roof and the rest of the building are protected by rated fire door or fire damper assemblies. Y N
j. If an attic exists, it is: NA
   1. Used for storage of flammables or combustibles Y N
   2. Used to house mechanical equipment Y N
   3. Protected by an automatic fire extinguishing system Y N

14. When the building share a common wall: NA
    a. Both buildings meet the minimum construction standards Y N
    b. The common wall serves as a fire separator Y N
    c. Openings in the common wall serve as a horizontal exit. Y N

15. The facility shares the building with another occupancy Y N
    (If "Yes," specify the occupancy)

    _____
    _____
    _____

16. A fire separation is provided between the facility and all other occupancies within the buildings Y N

17. Required fire separations are: NA
    a. Constructed for materials that have at
       least a 2-hour fire-resistive construction   Y   N

    b. Continuous through any concealed space
       from slab to floor or roof slab above   Y   N

    c. Continuous from exterior wall to
       exterior wall   Y   N

    d. Protected by openings by:
       1. Doors that have listed Class A or B
          swinging fire doors with positive
          latching devices   Y   N
       2. Doors that are kept closed or held
          open only by approved devices   Y   N
       3. Listed fusible link activated fire
          dampers (in ducts)   NA   Y   N
       4. Listed fire dampers that respond
          to installed fire detection systems
          (in ducts)   NA   Y   N

18. Undercuts, louvers, or transom exist in or in association with:
    a. Fire doors   NA   Y   N
    b. Smoke partition doors NA   Y   N
    c. Other corridor doors   Y   N

_____

_____

(Note: If any of the above items are answered "Yes," indicate if the undercuts, louvers, and transoms are sealed smoke-tight and of fire resistive construction equal to the door or wall in which installed. If these requirements are not satisfied, please explain and indicate the location.

_____

_____

Appendix C — Statement of Construction and Fire Protection 213

19. Smokestop partitions are
    a. Constructed of materials that have at least
       a 1-hour resistive rating   Y   N

    b. Continuous through any concealed space
       from floor slab to floor or roof slab above   Y   N

    c. Continuous from outside wall to outside wall   Y   N

    d. Protected at door openings by doors that are
       1. Hollow metal   Y   N
       2. At least 1-3/4 inch thick solid
          bonded wood core   Y   N
       3. Part of listed Class C fire door
          assemblies   Y   N
       4. Normally kept closed or held open
          only by approved devices   Y   N

    e. Protected at door openings by doors that
       have vision panels
       1. Of fixed wired glass set in approved
          steel or other approved metal frames   Y   N
       2. Other than described above   Y   N
          (specify differences)

          _____

          _____

    f. Protected by vision panels at openings   NA
       other than doors that are of:
       1. Fixed wire glass set in approved
          steel or other approved metal frames   Y   N
       2. Other (describe)   Y   N

          _____

          _____

    g. That have duct penetrations protected
       by approved smoke dampers that close
       upon activation of the: NA

1. Fire alarm   Y   N
2. Automatic fire extinguishing system Y   N
3. Smoke detector within duct   Y   N
4. Local smoke detector in the corridor Y   N

20. All openings into vertical shafts (stairwell, mechanical shaft, elevator and conveyor shaft, light and ventilation shaft) protected by at least a 1-hour fire resistive rated material, with at least the equivalent of Class B fire doors.   NA   Y   N

(Note: For buildings constructed after 1988, a 2-hour fire resistive rating is required for stairways in buildings 2 or more stories in height.)

21. Do any open stairways exist within the building?   Y   N
(Note: If answer is "Yes," state the location)

_____
_____
_____

22. Stairwell doors are: NA
    a. Self-closing   Y   N
    b. Equipped with positive latches   Y   N
    c. Provided with vision panels: Y   N
       1. Of wired glass   Y   N
       2. Set in approved steel or other   Y   N
          approved metal frames

23. There are chutes in this building   Y   N
(Note: If answer is "No," go on to question number 24.)

    a. All chutes are sealed by fire resistive construction and no longer in use   Y   N

Appendix C — Statement of Construction and Fire Protection

(Note: If answer is "No," complete the following. If answer is "Yes," go on to question 24.)

b. In buildings with less than four stories all chutes are enclosed with construction having at least a 1-hour fire resistive rating   NA   Y   N

c. In building with four or more stories all chutes are enclosed with construction having at least a 2-hour fire resistive rating   NA   Y   N

d. All chutes entrances are provided with positive latching, self-closing, Class B fire door assemblies   Y   N

e. Chute service doors open directly onto a corridor or exitway   Y   N

f. Chute service doors are located in a room enclosed with 1-hour noncombustible fire resistive construction   Y   N

g. All chutes are provided with automatic sprinkler protection   Y   N

h. All chute outlets are provided with approved doors that are self-closing with positive latches   Y   N

i. All chute outlets are routinely kept closed or are automatic?   Y   N

j. All chutes are vented through the roof   Y   N

24. At least two approved exits, remote from each other, are provided for each floor or fire section, with at least one exit being other than a horizontal exit? Y   N

(Note: If answer is "No," please explain.)

_____

_____

25. Exit Characteristics:
    a. The travel distance to an exit measures 100 feet or less from the entrance door of any room (for buildings protected throughout by an approved automatic fire extinguishing system, substitute 150 feet for 100 feet Y   N

    b. The travel distance between any point in a room and an exit exceeds 150 feet Y   N

    c. All means of egress are illuminated by at least 1.0 (one) footcandle at the floor Y   N

    d. Every exit and exit access is identified by a readily visible sign Y   N

    e. Exit and exit directional signs are continuously illuminated with a reliable light source connected to the emergency power system Y   N

26. Dead-end corridors of more than 30 feet exist within the structure Y   N

(Note: If answer is "Yes," state the location and the length in feet.)

_____

_____

_____

Appendix C — Statement of Construction and Fire Protection

27. Interior finish in exit hallways is:

    a. Class A (flame spread 0-25)   Y   N

    b. Class B (flame spread 26-75, specify location   Y   N

    c. Class C (flame spread 76-200, specify location)   Y   N

    d. Other (specify flame spread and location)   Y   N

28. Floor coverings installed after January 1, 1988 meet: NA

    a. Class I   Y   N

    b. Class II   Y   N

    c. Other (state location)   Y   N

29. All stairs serving as required means of egress have   NA

    a. At least one handrail, on either side going down   Y   N

    b. Handrails on both sides   Y   N

    c. Discharge to the outside at grade   Y   N

    d. Discharge through a protected passage to an approved exit discharge   Y   N

30. Are horizontal exits provided in the facility?   Y   N

(Note: If answer is "Yes," complete the following. If answer is "No," proceed to question number 31.)

a. Are these exits in walls or partitions that are of noncombustible material having at least a 2-hour fire resistive rating?   Y   N

b. Are these exits in walls or partitions that provide a separation that is continuous from the top floor (or roof) to the ground?   Y   N

c. Are these exits in walls or partitions that provide a separation that is continuous from exterior wall to exterior wall?   Y   N

d. Do these exits have doors that are routinely kept closed or are held open only by approved devices?   Y   N

31. Doors in fire separations, horizontal exits, stairways, and smokestop partitions are:

	Fire Separations	Horizontal Exits	Smokestop Partitions	Stairways
a. Always closed except when in use	[ ]	[ ]	[ ]	[ ]
b. Closed upon activation of fire alarm system	[ ]	[ ]	[ ]	[ ]
c. Closed upon activation of a complete automatic sprinkler system	[ ]	[ ]	[ ]	[ ]
d. Closed upon activation of a local smoke detection device	[ ]	[ ]	[ ]	[ ]
e. Closed upon activation of a fusible link	[ ]	[ ]	[ ]	[ ]

Appendix C — Statement of Construction and Fire Protection

f. Held upon by door stops or other such methods   [ ]   [ ]   [ ]   [ ]

32. Does this building have an electrically supervised, manually operated fire alarm system throughout?   Y   N

    (Note: If answer is "No," please explain) _____
    _____
    _____

33. If number 32 is answered "Yes," does the system transmit an alarm automatically to the fire department that serves the facility by: NA
    a. A direct connection   Y   N
    b. A municipal alarm auxiliary connection   Y   N
    c. A connection to an approved central station system   Y   N
    d. Other (explain) _____
    _____
    _____

34. Is there is automatic sprinkler protection, the system includes: NA
    a. A water flow alarm connection to the fire alarm system   Y   N
    b. Main control valve electrical supervision to sound an alarm at a constantly attended location   Y   N

35. Are the commercial cooking ranges and
    deep fat fryers provided with one of the
    following fixed automatic fire extin-
    guishing systems   NA
    a. Carbon dioxide   Y   N
    b. Dry chemical   Y   N
    c. Approved automatic sprinkler system   Y   N
    d. Foam water system   Y   N
    e. Other (explain) _____
    _____
    _____

(Note: If none of the above are answered "Yes," please explain.)
_____
_____
_____

36. Do the fire extinguishing systems for the
    cooking ranges and deep fat fryers
    protect the: NA
    a. Hoods   Y   N
    b. Ducts   Y   N
    c. Cooking surfaces   Y   N
    d. Grease removal devices   Y   N

Use this space for any additional comments relating to any questions answered in this section. When doing so, please refer to the question by number.
_____
_____
_____
_____
_____
_____

APPENDIX D
GLOSSARY

Abatement - a reduction or diminishing

A.B.M.A. - American Boiler Manufacturers Association

Absolute Humidity - the weight in grains of water vapor actually contained in 1 cubic foot of the air and moisture mixture

Absolute Pressure - atmospheric pressure added to gage pressure

Absolute Temperature - the theoretical temperature when all molecular motion of a substance stops; minus 460 degrees Fahrenheit

Access Flooring - a raised floor consisting of removable panels under which ductwork, wiring and pipe runs are installed

Acid Cleaning - a process in which dilute acid, used in tandem with a corrosion inhibitor, is applied to metal surfaces for removing foreign substances too firmly attached

Acoustical Ceiling - a ceiling composed of tiles having sound absorbing properties

Acoustical Tile - special tile for walls and ceilings made of mineral, wood, vegetable fibers, cork, or metal. Its purpose is to control sound volume, while providing cover.

Acutator - a device that converts a pneumatic or electric signal to force which produces movement

ADR - alternative dispute resolution; use of a mediator or other neutral third party instead of litigation to resolve a dispute

A.E.E. - Association of Energy Engineers

AHERA - Asbestos Hazard Emergency Response Act (1986); federal law requiring local education agencies to identify asbestos hazards and develop abatement plans

AHU - air handling unit

Air Changes - the number of times in an hour that a volume of air filling a room is exchanged

Algae - a form of plant life which causes fouling in water system piping; especially in cooling towers

Alkaline - a condition of liquid, opposite from acidic on the pH scale, which is represented by carbonate, bicarbonates, phosphates, silicates or hydroxides contained within it

Ambient Temperature - the temperature of the air immediately surrounding a device

Amplifier - a device that receives an input signal from an independent source and delivers an output signal that is related to, and generally greater than the input signal

Analog - a continuous range of values such as temperature, pressure, etc. (contrast with binary)

Anemometer - an instrument used for measuring air velocity

Anthracite coal - a dense coal known for its low volatility which enables the use of smaller combustion chambers than those needed to burn bituminous coal

Apron - a paved area, such as the juncture of a driveway with the street or with a garage entrance

Astragal - a molding or strip used to cover the joint where two doors meet

Atmospheric Pressure - the weight of the atmosphere measured in pounds per square inch

Atomization - the process of reducing a liquid into a fine spray

Auxiliary Device - a component, added to a control system, which when actuated by the output signal from one or more controllers produces a desired function

AWG - American wire gauge

Axial Fan - a device which discharges air parallel to the axis of its wheel

Backing Plate - a steel plate positioned behind a welding groove to confine the weldment and assure full penetration

Backwash - the backflow of water through the resin bed of a water softener during the cleaning process

Baffle - a structure or partition used for directing the flow of gasses or liquids

Bagasse - the dry pulp remaining from sugar cane after the juice has been extracted; a fuel used in boiler furnaces

Balanced Draft - a fixed ratio of incoming air to outgoing products of combustion

Balusters - upright supports of a balustrade rail

Balustrade - a row of balusters topped by a rail, edging a balcony or a staircase

Batt - insulation in the form of a blanket, rather than loose filling

Batten - Small thin strips covering joints between wider boards on exterior building surfaces

BDC - bottom dead center; when a piston is at the bottom of its stroke

Bearing Wall - a wall structure which supports floors and roofs

BHP - brake horsepower; the actual power produced by an engine

Biocide - a substance that is destructive to living organisms that is used in refrigeration systems by design

Bituminous Coal - a soft coal generally more volatile and requiring larger combustion chambers in which to burn than anthracite coal

Blister - a raised area on the surface of metal caused by overheating

Blowback - the difference in pressure between when a safety valve opens and closes

Blowdown - the removal of water from a boiler in lowering its chemical concentrations

B.O.C.A. - Building Officials and Code Administrators International, Inc.

Boiler Horsepower - the evaporation of 34.5 pounds of water per hour from a temperature of 212°F into dry saturated steam

Boiling Out - a process whereby an alkaline solution is boiled within a vessel to rid its interior of oil or grease

Boiling Point - the temperature at which a liquid is converted to a vapor corresponding to its pressure

B.O.M.A. - Building Owners and Managers Association

B.O.M.I - Building Owners and Managers Institute, International, Inc.

Boyles Law - a law of physics dealing with variations in gas volumes and pressures at constant temperatures

Braced Framing - construction technique using posts and cross-bracing for greater rigidity

Breeching - a large duct used for conveying gasses of combustion from a furnace to a stack

Brick Veneer - brick used as the outer surface of a framed wall

Bridging - small wood or metal pieces placed diagonally between floor joints

British Thermal Unit (Btu) - a unit measurement of heat; the amount of heat needed to raise the temperature of one pound of water, one degree Fahrenheit

Building Paper - heavy paper used in walls or roofs to damp-proof

Built-Up Roof - a roofing material applied in sealed, waterproof layers, where there is only a slight slope to the roof

Bus - vertical and horizontal metal bars which distribute line-side electrical power to branch circuits

Bushing - a removable sleeve inserted or screwed into an opening to limit its size

Butterfly Valve - a throttling valve consisting of a centrally hinged plate that can be opened partially or expose the full cross section of the pipe it feeds by maneuvering the valve handle through a quarter turn

BX - electrical cable wrapped in rubber with a flexible steel outer covering

Calorie - the quantity of heat needed to raise the temperature of one gram of water, one degree centigrade

Cantilever - a projecting beam or joist, not supported at one end, used to support an extension of a structure

Capillary Action - the capacity of a liquid to be drawn into small spaces

Carryover - a condition whereby water or chemical solids enter the discharge line of a steam boiler

Casing - the outer skin or enclosure forming the outside of an appliance

Cavitation - the formation of vapor pockets in a flowing liquid

CERCLA - Comprehensive Environmental Responsibility, Compensation and Liability Act (1976); also Superfund: federal law authorizing identification and remediation of abandoned hazardous waste sites

Cfc - Chlorofluorocarbon; chemical substance associated with depletion of Earth's ozone layer

Cfm - cubic feet per minute

CH Ratio - carbon-hydrogen ratio

Chassis - the frame or plate on which the components of a device are mounted

Chimney Effect - the tendency of air to rise within confined vertical passages when heated

Chlorination - the addition of the chemical chlorine to water

C I D - cubic inch displacement

Cistern - a tank to catch and store rain water

Closed Loop System (HVAC) - the arrangement of components to allow system feedback, e.g., a heating unit; valve and thermostat arranged so that each component affects the other and can react to it

Coagulation - the initial aggregation of finely suspended matter by the addition of floc forming chemical or biological process

Coaxial Cable - cable that consists or a tubular conductor surrounding a central conductor held in place by insulating material. Used for transmitting high frequency signals.

Coefficient of Heat Transmission (U) - the amount of heat measured in Btu's transmitted through materials over time; the heat transmitted in one hour per square foot per degree difference between the air inside and outside of a wall, floor or ceiling

Coefficient or Performance (COP) - the ratio of work performed to the energy used in performing it

Combustion Efficiency - the ratio of the heat released from a fuel as it burns, to its heat content

Comfort Zone - a range of temperature and humidity combinations within which the average adult feels comfortable

Concentrations - the number of times that dissolved solids increase in a body of water as a one-to-one multiple of the original amount due to the evaporation process

Condensation - the process of returning a vapor back to its liquid state through the extraction of latent heat

Conduit - a pipe, tube or tray in which electrical wires are run and protected

Contingency Planning - a process which anticipates and prescribes corrective action to be taken in the event of unforeseen circumstances and emergency situations

Continuous Blowdown - a process whereby solids concentrations within a body of water are controlled through the constant removal and replacement of the water

Convection - a process of heat transfer resulting from movement within fluids due to the relative density of its warmer and cooler parts

Corbel - a horizontal projection from a wall, forming a ledge or supporting a structure above it

Cornice - a horizontal projection at the top of a wall or under the overhanding part of the roof

Corrosion - the wasting away of metals due to physical contact with oxygen, carbon-dioxide or acid

Counterflashing - a downward turned flashing which overlaps an upward turned flashing used for protecting against the entry of water into a structure

Counterflow - a method of heat exchange that brings the coldest portion of one moving fluid into contact with the warmest portion of another

CPU - central processing unit; that portion of a computer which contains the arithmetic and logic functions which process programmed instructions

Crude Oil - unrefined petroleum in its natural state as it comes from the ground

Curing Compound - a liquid which is sprayed onto new concrete to prevent premature dehydration

Damper - a mechanism used to create a variable resistance within a gas or air passage in order to regulate its rate of flow

DB - decibel unit for describing noise level

Deaeration - the removal of entrained air from a liquid

Dedicated - set apart or committed to a definite use (wiring, conduit, etc.); addressed to a definite task (signals)

Degree Day - a unit representing one degree of difference from a standard temperature in the average temperature of one day, used to determine fuel requirements

Degree of Superheat - the difference between the saturation temperature of a vapor and its actual temperature at a given pressure

Demineralization - deionization; the removal of ionizable salts from solution

Desiccant - a drying agent such as silica gel or activated alumina that is used to absorb and hold moisture

Dew Point Temperature - the lowest temperature that air can be without its water vapor condensing

Direct Current - an electrical current which flows in only one direction

Door Buck - the rough frame of a door

Dormer - the projecting frame of a recess in a sloping roof

Double Glazing - an insulating window pane formed of two thicknesses of glass with a sealed air space between them

DOT - U.S. Department of Transportation; enforces regulations governing the transport of hazardous materials

DPDT Switch - double pole double throw switch

DRE - destruction and removal efficiency; measure of the effectiveness of

Appendix D — Glossary

incineration in removing contaminants from waste materials

Dry Bulb Temperature - the temperature of the air as measured on a thermometer

Duct Furnace - a furnace located in the ducting of an air distribution system to supply warm air for heating

DWV - drain, vent, waste pipes

Economizer - a heat recovery device that utilizes waste heat for preheating fluids

Efflorescence - chemical salt residue deposited on the face of masonry caused by the infiltration of water into a structure

Effluent - treated sewage from a septic tank or sewage treatment plant

Electric Ignition - ignition of a pilot or burner by an electric spark generated by a transformer

Electrolysis - a chemical reaction between two substances prompted by the flow of electricity at their point of contact

Electrostatic Precipitator - an electrically charged device used for removing dust particles from an air stream

Elevator Recall - Overriding control of normal elevator operation by the fire safety system. Elevators are automatically directed to ground floor the instant an alarm is sounded, and placed under full control of the Fire Department.

EMF - electromotive force; voltage

Enthalpy - the actual or total heat contained in a substance. It is actually calculated from a base. In refrigeration work the base temperature is accepted as –40(degrees F) –40(degrees C).

Enthalpy Switchover - automatic switching or regulation of outside air and return air dampers. Total heat content of inside and outside air is compared before selecting either inside or outside or a mixture of the two for ventilating, which will require the least amount of refrigeration, humidification or dehumidification.

Entrainment - the inclusion of water or solids in steam, usually due to the violent action of the boiler process

EPA - U.S. Environment Protection Agency; primary federal agency responsible for enforcement of federal laws protecting the environment; also referred to as the Agency.

Evacuation - the removal of air and moisture from a refrigeration system

Evaporation - the transformation of a liquid into its vapor state through the application of latent heat

Excess Air - the air supplied for the combustion process in excess of that theoretically needed for complete oxidation

Face and Bypass - duct and damper system which directs air through (face) or around (bypass) heating or cooling coils

Fail Safe Control - a device that opens an electric circuit when the sensing element senses an abnormal condition

Farad - the unit of electrical capacity of a capacitor

Fascia - a flat horizontal member of a cornice placed in a vertical position

Feedwater Treatment - the conditioning of water with chemicals to establish wanted characteristics

FIFRA - Federal Insecticide, Fungicide and Rodenticide Act (1972, 1988); federal law mandating toxicity testing and registration of pesticides.

Fire Separation Wall - a wall dividing two sections of a building used to prevent the spread of fire

Flame Rod - a metal or ceramic rod extending into a flame which functions as an electrode in a flame detection circuit

Flame Safeguard System - the equipment and circuitry used to provide safe control of burner operation

Flame Simulator - a device used as a substitute for the presence of flame to test a flame detection circuit

Flame Spread Rating - the measure of how fast fire will spread across the surface of a material once it is ignited

Flashback - the backward movement of a flame through a burner nozzle

Flash Point - the minimum temperature necessary for a volatile vapor to momentarily ignite

Floating Action - Movement of the controlled device either toward its open or its closed position until the controller is satisfied, or until the controlled device reaches the end of its travel or until a corrective movement in the opposite direction is required. Generally there is a neutral zone in which no motion of the controlled device is required by the controller.

Flocculant - a chemical used to bridge together previously coagulated particles

Fluoridation - the addition of the chemical fluoride to water

Flow Rate - the amount of fluid passing a given point during a specified period of time

Flue - a pipe or conduit used for conveying combustion exhaust fumes to the atmosphere

Appendix D —Glossary

Foaming - the continuous formation of bubbles having a high surface tension which are hard to disengage from a surface

Forced Draft - the process of moving air mechanically by pushing or drawing it through a combustion chamber with fans or blowers

Fouling - the accumulation of refuse in gas passages or on heat absorbing surfaces which results in undesirable restrictions to flow

Furring - Thin wood, or metal applied to a wall to level the surface for lathing, boarding, or plastering, to create an insulating air space, and to dampproof the wall.

Fusible Plug - a safety plug used in refrigerant containers that melts at a high temperature to prevent excessive pressure from bursting the container

Gage Pressure - absolute pressure minus atmospheric pressure

Gauge Manifold - a manifold that holds both the pressure and compound gauges, the valves that control the flow of liquids

GPG - grains per gallon. one grain per gallon equals 17.1 ppm

GPM - gallons per minute

Grain - a unit of weight equal to 1/7000 lb (0.06480 g), which is used to indicate the amount of moisture in the air

Gravity Feed - the transfer of a liquid from a source to an outlet using only the force of gravity to induce flow

Grooving - a form of corrosion wherein a groove is formed along the length of tubes or shells

Ground - an electrical connection made between any structure or object and the earth

Halogens - chlorine, iodine, bromine or fluorine

Halide Torch - a device which uses an open flame for detecting refrigerant leaks

Hardness - a term used to describe the calcium and magnesium content of water

Header - a manifold to which many branch lines are connected

Heat Exchanger - a device used to transfer heat from one medium to another

Heat Gain, Total - the sum total of sensible plus latent heat gain from ventilation and infiltration of outside air (convection), heat conduction through walls and roofs, solar radiation and heat generated by people, lights and machinery.

Heating Surface - that portion of a heat exchange device which is exposed to the heat source and transfers heat to the heated medium

Hermetic Compressor - a unit wherein a compressor and its driving motor are contained in a single, sealed housing

Hertz - (Hz) one cycle per second

Hg (mercury) - a silver-white, heavy, liquid metal. It is the only metal that is a liquid at room temperature.

High Fire - the firing rate at which a burner consumes the most fuel this producing the most heat

High Limit - the maximum value at which a controller is set that if exceeded causes the shut down of a system

High Side - the part of the refrigeration system that contains the high-pressure refrigerant. Also refers to the outdoor unit, which consists of the motor, compressor, outdoor coil, and receiver mounted on one base.

High-Side Charging - The process of introducing liquid refrigerant into the high side of a refrigeration system. The acceptable manner for placing the refrigerant into the system.

High Temperature Boiler - a boiler which produces hot water at pressures exceeding 160 p.s.i. or at temperatures exceeding 250 degrees Fahrenheit

Horsepower - (hp) - a unit of power equal to 550 foot pounds per second, 33,000 foot pounds per minute or 746 watts.

HRS - Hazard Ranking System; system used to rank NPL sites in terms of degree of contamination and urgency for remediation

HRT - horizontal return tubular boiler

HSWA - Hazardous and Solid Waste Amendments; 1984 amendments to RCRA establishing a timetable for landfill bans and more stringent requirements for USTs.

Humidistat - a control device which responds to changes in the humidity of air

HWTC - Hazardous Waste Treatment Council; Washington-based trade association of more than 60 treatment and disposal firms

Hydrostatic Test - a procedure in which water is used to determine the integrity of pressure vessels

IC - internal combustion

ID - inside diameter

Ignition Temperature - the minimum temperature at which the burning process can begin for a given fuel source

Induced Draft Fan - a fan or blower located in the breeching or gas passages that produces a negative pressure in the combustion chamber causing air to be drawn through it

Inerts - non-combustible particles found in fuel

Interlock - a sensor or switch which monitors the status of a required condition which causes a programmed action to occur when the condition becomes inappropriate

Interrupting Capacity - a rating given to a piece of electrical equipment. The rating represents the maximum short circuit conditions under which the equipment will not disintegrate. This rating must be equal to or greater than the short circuit current available at the particular building location.

I/O Devices - a device used to convey information to or from a computer, e.g., a keyboard or printer; input/output

Ion Exchange - a process for removing impurities from water on the atomic level through the selective repositioning of electrons.

Ionization Detector - Detects fire in its very beginning stages, before smoke, flame or appreciable heat is present by detecting invisible products of combustion (referred to as the incipient stage).

Interface - a shared boundary. An interface might be a hardware component to link two devices or it might be a portion of storage or registers accessed by two or more computer programs.

R Drop - voltage drop across a resistance in an electrical circuit

Joist - a small rectangular sectional member arranged parallel from wall to wall in a building, or resting on beams or girders. They support a floor on the laths or furring strips of a ceiling.

Jumper - a short length of wire used to bypass all or part of an electrical circuit

Kelvin Scale - a temperature scale incremented in centigrade that begins at absolute zero (–273C)

Kiln-Dried - artificial drying of lumber, superior to most lumber that is air dried

King-Post - the middle post of a truss

Knockout - portal designed into the side of an electrical box or metal cabinet that can be easily removed to accommodate wires or piping

KVA - kilovolt amperes

Lally Column - a steel tube sometimes filled with concrete, used to support girders or other front beams

Latent Heat of Condensation - the heat extracted from a vapor in changing it to a liquid with no change in temperature

Latent Heat of Evaporation - the heat added to a liquid in changing it to a vapor with no change in temperature

Latent Heat of Fusion - the heat added to a solid in changing it to a liquid with no change in temperature

Leaching Bed - tiles in the trenches carrying treated wastes from septic tanks

Lighting Arrester - a device located in an electrical circuit to protect it from the effects of lighting

Lithium Bromide - a chemical having a high affinity for water used as a catalyst in absorption refrigeration systems

Load Shedding - Shutting down non-critical electrical equipment to a pre-selected level, when a peak electric use period approaches. This is done to prevent paying an excessive electric rate which is based on highest electric usage during a billing period.

Locked Rotor - a test in which a motor's rotor is locked in place and rated voltage is applied

Locked Rotor Amps - the amperage which is apparent in a live circuit of a motor driven device when the rotor is not moving

Louver - an opening with horizontal slats to permit passage of air, but excluding rain, sunlight and view

Low Limit - the minimum value at which a controller is set that if dropped below will result in a shut down of the system

Low Pressure Boiler - a steam boiler whose maximum allowable working pressure does not exceed 15 p.s.i.

Low Side - those parts of a refrigeration system in which the refrigerant pressure corresponds to the evaporating coil pressure

Low-Side Charging - the process of introducing refrigerants into the low side of the system; usually reserved for the addition of a small amount of refrigerant after repairs

Low Voltage Relay (Integrated Control Center) - a relay which initiates load shutdown whenever line voltage is less than 83% of the normal line voltage

Low Water Cutoff - a mechanism used for shutting off the supply of fuel to a furnace when a boiler's water level falls to a dangerously low level

LPG - liquified petroleum gas

Makeup Water - water added to a system to replace that which was lost during operation due to leaks, consumption, blow down and evaporation

Manometer - a U-shaped tube used for measuring pressure differences in air passages

Manual Reset - the operation required after a system undergoes a safety shutdown before it can be put back into service

Master - an instrument whose variable output is used to change the set point of a submaster controller. The master may be a humidistat, pressure controller, manual switch, transmitter, thermostat, etc.

MAWP - maximum allowable working pressure

Measured Variable - the uncontrolled variable such as temperature, relative humidity or pressure measured by a sensing element

Mega - one million times

Megohmeter - an instrument used for evaluating the resistance values of electrical wire coverings

MHO - a unit measurement of electrical conductance

Micron - one millionth of a meter; 1.25,400 in.

Mixing Valve - a three-way valve having two inlets and one outlet designed specifically for mixing fluids

Modulating Fire - varying the firing rate with the load thereby decreasing the on-off cycling of burners

Mullion - slender framing which divides the lights or panes or windows

N.E.C. - National Electrical Code

Natural Circulation - the circulation of fluids resulting from differences in their density

NC - normally closed; a relay contact which is closed when the relay coil is not energized

NCP - National Contingency Plan (National Oil and Hazardous Substances Pollution Contingency Plan); regulations promulgated by EPA to implement CERCLA and Sec. 311 of CWA.

Newel - the upright post or the upright formed by the inner or smaller ends of steps about which steps of a circular staircase wind. In a straight flight staircase, the principal post at the floor or the secondary post at a landing.

Nitrogen Blanket - a technique used whereby the air space above a body of water in a vessel is filled with nitrogen to keep oxygen from coming into contact with its metal surfaces

NO - normally open; a relay contact which is open when the relay coil is not energized

Nominal Dimension - an approximate dimension; a conventional size

Non-Condensable Gas - any gas, usually in a refrigeration system, that cannot be condensed at the temperature and pressure at which the refrigerant will condense, and therefore requires a higher head pressure

Normally Closed - applies to a controlled device that closes when the signal applied to it is removed; (closing a pneumatic device stops the flow of the control agent; closing an electrical device allows an electrical current to flow)

Normally Open - applies to a controlled device that opens when the signal applied to it is removed; (opening a pneumatic device allows the flow of the control agent); (opening an electrical device stops the flow of electrical current)

NPDES - National Pollutant Discharge Elimination System; federal permitting system required by EPA for hazardous effluents

NPL - National Priorities List; official list of hazardous waste sites to be addressed by CERCLA

NPT - national pipe thread

NRC - Nuclear Regulatory Commission; federal body overseeing the operation of nuclear power plants and other installations

OD - outside diameter

OHM - a unit measurement of electrical resistance

OHM's Law - a mathematical relationship between voltage, current, and resistance in an electrical circuit. It is stated as follows: voltage (E) = amperes (I) X ohms (R).

Oil Binding - a condition in which a layer of oil on top of liquid refrigerant may prevent it from evaporating at its normal pressure and temperature

One Pipe System - a system in which one pipe serves as both the supply and return main

Open Circuit - an electrical circuit in which the current path has been interrupted or broken

Orsat - a device used to analyze gasses by absorption into chemical solutions

OSHA - Occupational Safety and Health Administration; oversees and regulates workplace health and safety

OTA - Office of Technology Assessment; federal office responsible to the U.S. Congress for carrying out research and identifying policy alternatives on technology related issues

Overload Alarm Circuit - (Integrated Control Center) - used to sound an alarm, light a pilot light locally or send a communique to a central building automation console. Is usually mechanically linked to the overload contacts of the starter in an integrated control center.

Overload Protector - a safety device designed to stop motors when overload conditions exist

Overshoot - the greatest amount a controlled variable deviates from its desired value before stabilizing, after a change of input

Oxygen Scavenger - a chemical treatment such as sulfite or hydrazine used for releasing dissolved oxygen from water

Package Boiler - one that is shipped from the assembly plant completely equipped with all the apparatus needed for its operation

Panel Heating - a method whereby interior spaces are heated by pipe coils located within walls, floors and ceilings

Parging - a rough coat of mortar applied over a masonry wall as protection of finish; may also serve as a base for an asphaltic waterproofing compound below grade

PCB - polychlorinated biphenyl; a pathogenic and teratogenic industrial compound used as a heat-transfer agent; PCBs may accumulate in human or animal tissue.

PE - pneumatic/electric relay

Perfect Combustion - the complete oxidation of a fuel using no excess air in the combustion process

Peripheral Device - a hardware item forming part of a computer system that is not directly connected to but supports the processor

PF - power factor

pH - a value that indicates the intensity of the alkalinity or acidity of a solution

Pilaster - a projection or the foundation wall used to light off a main burner

Pitot Tube - a device that measures air velocity

Plenum Chamber - a compartment to which ducts are connected enabling the distribution of air to more than one area

Pointing - treatment of joints in masonry by filling with mortar to improve appearance or protect against weather

Poly-Phase Motor - an electric motor driven by currents out of phase from circuit to circuit

Positive Displacement - an action wherein the total amount of a fluid being transferred by a mechanical device is accomplished without leakage or back siphonage

Potentiometer - a variable resistor in an electrical circuit

Power Factor - an efficiency value assigned to electrical circuits based on a comparison of its true and apparent power characteristics

PPM - parts per million

Precipitation - the removal of constituents from water by chemical means; condensation of water vapor from clouds

Pressure Limiter - a device that remains closed until a predetermined pressure is reached and then opens to release fluid to another part of the system or opens an electrical circuit

Pressure Regulator - a mechanism used to maintain a constant pressure within a feeder line regardless of fluctuations above the setting in the supply line

Primary Air - combustion air which is introduced into a furnace with the fuel

Priming - the discharge of water particles into a steam line

Products of Combustion - any gas or solid remaining after the burning of a fuel

PRV - pressure reducing valve

PSI - pounds per square inch

Psychrometric Chart - a graph which depicts the relationship between the pressure, temperature and moisture content of air

Purge - eliminating a fluid from a pipe or chamber by flushing it out with another fluid

PVC - polyvinyl chloride

Raceway - any method commonly used and accepted for running electrical wires or cable or pneumatic lines within a building. Raceway may be exposed or concealed in floors, walls, ceiling plenums, or may be buried or installed in outdoor locations.

Radiation Loss - the loss of heat from an object to the surrounding air

RCRA - Resources Conservation and Recovery Act (1980); regulates management and disposal of hazardous materials and wastes currently being generated, treated, disposed or distributed.

Rectification - the conversion of alternating current to direct current

Appendix D — Glossary

Reed Values - a piece of thin, flat, tempered steel plate fastened to the valve plate

Refractory - heat resistant material used to line furnaces, ovens, and incinerators

Refrigeration - the removal of heat from an area where it is not wanted to one that is not objectionable

Register - the grill work or damper through which air is introduced

Reinforced Concrete - concrete strengthened with wire or metal bars

Relay - an electromechanical device having a coil which, when energized and de-energized, opens and closes sets of electrical contacts

Relief Valve - a device used to relieve excess pressure from liquid filled pressure vessels, pipes and hot water boilers

Reverse Acting - the output signal changing in the opposite direction the controlled or measured variable changes. Example: an increase in the controlled or measured variable results in a decreased output signal.

RH - relative humidity

Ringelmann Chart - a comparator of smoke density comprised of rectangular grids filled with black lines of various widths on a white background

RMS - root mean square

RPM - revolutions per minute

ROD - Record of Decision; the officially designated remedy chosen by the government for remediation of an NPL site

Running Current - the amperage draw noted when a motor is running at its rated speed

Safety Factor - the ratio of extra strength or capacity to the calculated requirements to ensure freedom from breakdown and ample capacity

Safety Plug - a device that releases the contents of a container to prevent rupturing when unsafe pressures or temperatures exist

Safety Valve - A quick-opening safety valve used for the fast relief of excessive pressure in a container

Sail Switch - a switch attached to a sail shaped paddle inserted into an air stream which is activated when the air stream striking the sail reaches a pre-established velocity

SARA - Superfund Amendments and Reauthorization Act (1986); federal law reauthorizing and expanding the jurisdiction of CERCLA.

SARA Title III - part of SARA mandating public disclosure of chemical information and development of emergency response plans

Saturated Steam - dry steam which has reached the temperature corresponding to its pressure

SBI - Steel Boiler Institute

Scale - a hard coating or layer of chemical materials on internal surfaces of pressure vessels, piping and fluid passages

Schrader Valve - a spring-loaded valve that permits fluid to flow in only one direction when the center pin is depressed

Sedimentation - settling out of particles from suspension in water

Sensible Heat - heat which changes the temperature but not the state of a substance

Sensible Heat Ratio - the percentage of total heat removed that is sensible heat. It is usually expressed as a decimal and is the quotient of sensible heat removed divided by total heat removed.

Set Point - a pre-determined value to which a device is adjusted that, when reached, causes it to perform its intended function

Shiplap - boards with rabbeted edged overlapping

Short Circuit - an unintentional connection between two points in an electrical circuit resulting in an abnormal flow of current

Shunt Trip - a coil mechanism which trips the breaker in response to an external voltage potential being applied to the coil

Slave - an instrument using the signal of a master unit to expand the master unit capabilities

Sling Psychrometer - a device having a wet and a dry bulb thermometer which measures relative humidity when moved rapidly through the air

Slugging - a condition in which a quantity of liquid enters the compressor cylinder, causing a hammering noise

SNG - synthetic natural gas

Spalling - deterioration of materials evidenced by flaking of their surfaces

Specific Gravity - the ratio of the wright of any substance to the same volume of a standard substance at the same temperature

Specific Heat - a measure of the heat required in Btu's to raise the temperature to one pound of a substance, one degree Fahrenheit; the specific heat of water is 1.0

Split-Phase Motor - a motor with two windings. Both windings are used in the starting of the motor. One is disconnected by a centrifugal

Appendix D — Glossary

switch after a given speed is reached by the motor. The motor then operates on only one winding

SSU - Seconds Saybolt Universal

Standard Atmosphere - a condition existing when the air is at 14.7 psia of pressure, 68 (degrees F) 20 (degrees D) temperature, and 36% relative humidity

Static Head - the pressure exerted by the weight of a fluid in a vertical column

Static Pressure - the force exerted per unit area by a gas or liquid measured at right angles to the direction of flow

Stratification of Air - a condition of the air when little or no movement is evident

Sump - a container, compartment or reservoir used as a drain or receptacle for fluids

Superheat - a temperature increase above the saturation temperature or above the boiling point

Swale - a wide shallow depression in the ground to form a channel for storm water drainage

TDC - top dead center; when a piston is at the top of its stroke

TDS - total dissolved solids

Tensile Strength - the capacity of a material to withstand being stretched

Terrazzo - a material used for poured floors consisting of concrete mixed with marble chips

Tertiary Air - air supplied to a combustion chamber to supplement primary and secondary air

Therm - a symbol used in the industry representing 100,000 Btu's

Thermistor - a solid state device whose electrical resistance varies with temperature

Thermocouple - a mechanism comprised of two electrical conductors made of different metals which are joined at a point which, when heated, produces an electrical voltage having the value that is directly proportional to the temperature of the heat being applied

Thermopile - a battery of thermocouples connected in series

Thermostatic Expansion Valve - a control device operated by the pressure and temperature of an evaporator which meters the flow of refrigerant to its coil

Three-Mode Switching - three-state switching operations such as fast-slow-stop, summer-winter-auto, day-night-auto, etc.

Ton of Refrigeration - the heat required to melt a 1-ton block of ice in 24 hours. 288,000 Btu's or 12,000 Btu's per hour

Transducer - an instrument that converts an input signal into an output signal, usually of another form, e.g., electrical input to pneumatic output

Two Position Valve - a valve which is either fully open or fully closed, having no positions in between

Undercarpet Wiring System - flat insulated wiring designed for running circuits under carpeting where access to wire chases under the floor are not available

Uninterruptible Power Supply - a separate source of electricity used to maintain continuity of electrical power to a device or system when the normal supply is interrupted

Unitary System - a combination heating and cooling system that is factory assembled in one package and is usually designed for conditioning one room or space

Vacuum - any pressure less than that of the surrounding atmosphere

Vapor Retarder - a barrier constructed of materials which retard the capillary action of water into building structures

VAR - volt-ampere reactive

VAV - variable air volume

Velocimeter - an instrument used to measure the speed of moving air

Venturi - a short tube designed with a constructed throat that increases the velocity of fluids passed through it

Viscosity - a measure of a fluid's resistance to flow

Wainscoting - the lower three or four feet of an interior wall when lined with paneling, tile or other material different from the rest of the wall

Weep Hole - a small hole in a wall which permits water to drain off

Wet Bulb Temperature - the lowest temperature that can be attained by an object that is wet and exposed to moving air

Work Relay - a device using a current sensing principle to detect the level of work a particular motor is doing. Control contacts of the device can activate a local indicator or can be tied into a central building automation system's console. The device is also equipped with an analog output.

Working Range - the desired controlled or measured variable values over which a system operates

APPENDIX E
TABLES

LIST OF TABLES

AIR
1. Btu Required for Heating Air
2. Composition of Air
3. Atmospheric Pressure per Square Inch
4. Approximate Air Needs of Pneumatic Tools
5. Average Absolute Atmospheric Pressure

CONSTRUCTION
6. Building Design Loads
7. Concrete Curing Methods
8. Concrete for Walls
9. Earth Excavation Factors
10. Lumber Sizes in Inches

CONVERSIONS
11. Weight
12. Temperature Conversion
13. Length and Area
14. Horsepower Equivalent
15. Approximate Metric Equivalents
16. Miscellaneous Measures

ELECTRIC
17. Appliance Energy Requirements

18. Alternating Current Calculations
19. Kilowatt Conversion Factors
20. Effects of Electric Current on Humans
21. Equivalent Electrical Units
22. United States Power Characteristics
23. Induction Motor Synchronous Speeds

FORMULAE
24. Geometric Formulas

HEAT
25. Heat Equivalents
26. Heat Generated by Appliances
27. Heat Loss from Hot Water Piping
28. Heat Loss from Bare Steel Piping
29. Specific Heat of Common Substances
30. Heat Content of Common Fuels
31. Typical Fuel Oil Analysis

MISCELLANEOUS
32. Cylindrical Tank Capacity
33. Metric Capacity
34. Metric Length
35. Metric Weight
36. Weights of Metals
37. Ultimate Bending Strength
38. Recommended Lighting Levels

STEAM
39. Steam Pressure Temperature Relationship
40. Steam Trap Selection

TEMPERATURE
41. Color Scale of Temperature
42. Centigrade/Fahrenheit Scale
43. Melting Points of Common Substances
44. Temperatures of Waste Heat Gases

WATER
45. Water Equivalents
46. Supply Line Sizes for Common Fixtures
47. Pressure of Water
48. Water Requirements of Common Fixtures
49. Blowdown Flow Rates — 3-Inch Pipe
50. Tank Capacities per Foot of Depth
51. Efficiency Loss Due to Scale

AIR

Table 1. Btu Required for Heating Air (1 cubic foot of air)
External Temperature of air in room

Temp	60	70	80	90
0	1.234	1.439	1.645	1.851
10	1.007	1.208	1.409	1.611
20	0.787	0.984	1.181	1.378
30	0.578	.0770	0.963	1.155
40	0.376	0.564	0.752	0.940
50	0.184	0.367	0.551	0.735
60		0.179	0.359	0.538
70			0.175	0.350

Table 2. Composition of Air

Nitrogen	78%
Oxygen	21%
Argon	0.96%
Carbon Dioxide & other gases	0.04%

Table 3. Atmospheric Pressure per Square Inch

Barometer (ins. of mercury)	Pressure (lbs. per sq. in.)
28.00	13.75
28.25	13.88
28.50	14.00
28.75	14.12
29.00	14.24
29.25	14.37
29.50	14.49
29.75	14.61
29.921	14.696
30.00	14.74
30.75	14.86
30.50	14.98
30.75	15.10
31.00	15.23

Table 4. Approximate Air Needs of Pneumatic Tools, CFM

Tool	CFM
Grinders, 6- and 8-in. diameter wheels	50
2- and 2-1/2-in. diameter wheels	14-20
File and burr machines	18
Rotary sanders, 9-in. diameter pads	55
Sand rammers and tampers:	
1 X 4-in. cylinder	25
1-1/4 X 5-in. cylinder	28
1-1/2 X 6-in. cylinder	39
Chipping hammers, 10-13 lb	28-30
2-4 lb	12
Nut setters to 5/16 in., 8 lb	20
1/2 to 3/4 in., 18 lb	30
Paint spray	2-20
Plug Drills	40-50
Riveters, 3/32- to 1/8-in. rivets	12
Rivet busters	35-39

Appendix E — Tables

Steel drills, rotary motors:
   to 1/4 in., weighing 1-1/4 to 4 lb 18-20
   1/4 to 3/8 in., weighing 6-8 lb 20-40
   1/2 to 3/4 in., weighing 9-14 lb 70
   7/8 to 1 in., weighing 25 lb 80
Wood borers to 1-in. diameter,
   weighing 14 lb 40

Table. 5 Average Absolute Atmospheric Pressure

Altitude in feet reference to sea level	Inches of Mercury (in. Hg)	Pounds per sq. in. absolute (psia)
− 1,000	31.00	15.2
− 500	30.50	15.0
sea level 0	29.92	14.7
+ 500	29.39	14.4
+ 1,000	28.87	14.2
+ 1,500	28.33	13.9
+ 2,000	27.82	13.7
+ 3,000	26.81	13.2
+ 4,000	25.85	12.7
+ 5,000	24.90	12.2
+ 6,000	23.98	11.7
+ 7,000	23.10	11.3
+ 8,000	22.22	10.8
+ 9,000	21.39	10.5
+ 10,000	20.58	10.1

CONSTRUCTION

Table 6. Building Design Loads

Occupancy or Use	Live Load Lbs. per Sq. Ft.
Apartment houses:	
Private apartments	40
Public stairways	100

Table 6. Building Design Loads (continued)

Occupancy or Use	Live Load Lbs. per Sq. Ft.
Assembly halls:	
Fixed seats	60
Movable seats	100
Corridors:	
First Floor	100
(Other floors, same as occupancy served except as indicated.)	
Dining rooms	100
Dwellings	40
Loft Buildings	125
Sidewalks	250
Stairways	100

Table 7. Concrete Curing Methods

Method	Advantage	Disadvantage
Wetting	Excellent results if constantly kept wet	Difficult on vertical walls
Straw	Insulator in winter	Can dry out, blow away, or burn
Curing Compounds	Easy to apply. Inexpensive	Sprayer needed. Can allow concrete to get too hot
Waterproof Paper	Prevents drying	Cost can be excessive
Plastic Film	Absolutely watertight	Must be weighed down

Table 8. Concrete for Walls
(Per 100 Square Feet Wall)

Thickness	Cubic Feet Required	Cubic Yards Required
4"	33.3	1.24
6"	50.0	1.85
8"	66.7	2.47
10"	83.3	3.09
12"	100.0	3.70

Table 9. Earth Excavation Factors

Depth	Cubic Yards per Square Foot	Depth	Cubic Yards per Square Foot
2"	.006	4'-6"	.167
4"	.012	5'-0"	.185
6"	.018	5'-6"	.204
8"	.025	6'-0"	.222
10"	.031	6'-6"	.241
1'-0"	.037	7'-0"	.259
1'-6"	.056	7'-6"	.278
2'-0"	.074	8'-0"	.296
2'-6"	.093	8'-6"	.314
3'-0"	.111	9'-0"	.332
3'-6"	.130	9'-6"	.350
4'-0"	.148	10'-0"	.369

Figure 10. Lumber Sizes in Inches

Nominal	Seasonal	Nominal	Seasonal
1 x 4	3/4 x 3-1/2	2 x 10	1-1/2 x 9-1/4
1 x 6	3/4 x 5-1/5	2 x 12	1-1/2 x 11-1/4
1 x 8	3/4 x 7-1/4	4 x 4	3-1/2 x 3-1/2
1 x 10	3/4 x 9-1/4	4 x 6	3-1/2 x 5-1/2
1 x 12	3/4 x 11-1/4	4 x 8	3-1/2 x 7-1/4
2 x 4	1-1/2 x 3-1/2	4 x 10	3-1/2 x 9-1/4
2 x 6	1-1/2 x 5-1/2	4 x 12	3-1/2 x 11-1/4
2 x 8	1-1/2 x 7-1/4		

Conversions

Table 11. Weight
- 1 Pound .... 0.454 Kilogram
- 1 Short Ton (2000 lbs) .... 907 Kilograms
- 1 Long Ton (2240 lbs) .... 1016 Kilograms
- 1 Gram .... 15.432 Grains
- 1 Kilogram .... 2.205 Pounds
- 1 Metric Ton .... 2205 Pounds

Table 12. Temperature Conversions

$°F = 9/5°C + 32$
$°F = °R - 459.58$
$°K = °C + 273.16$
$°R = °F + 459.58$
$°C = 5/9 (°F-32)$
$°K = 5/9°R$

Table 13. Length and Area

1 statute mile (mi)	=	5280 feet
	=	1.609 kilometers
1 foot (ft)	=	12 inches
	=	30.48 centimeters
1 inch (in)	=	25.40 millimeters
100 ft per min	=	0.508 meter per sec
1 square foot	=	144 sq inches
	=	0.0929 sq meter
1 square inch	=	6.45 sq centimeters
1 kilometer (km)	=	1000 meters
	=	0.621 statute mile
1 meter (m)	=	100 centimeters (cm)
	=	1000 millimeters (mm)
	=	1.094 yards
	=	3.281 feet
	=	39.37 inches
1 micron	=	0.001 millimeter
	=	0.000039 inch
1 meter per sec	=	196.9 ft per min

Appendix E — Tables

Table 14. Horsepower Equivalent
 1 HP = 33,000 ft. lb. per min.
 1 HP = 550 ft. lb. per sec.
 1 HP = 2,546 B.t.u. per hr.
 1 HP = 42.2 B.t.u. per min.
 1 HP = .71 B.t.u. per sec.
 1 HP = 746 watts

Table 15. Approximate Metric Equivalents

1 Decimetre	=	4 inches
1 Meter	=	1.1 yards
1 Hectare	=	2-1/2 acres
1 Litre	=	1.06 qt.
1 Kilogramme	=	2.2 lb.
1 Metric Ton	=	2,200 lb.

Table 16. Miscellaneous Measures

ANGLES OR ARCS

60 seconds (")	=	1 minute
60 minutes (')	=	1 degree
90 degrees (°)	=	1 rt. angle or quadrant
360 degrees	=	1 circle

AVOIRDUPOIS WEIGHT

437.5 GRAINS (gr.)	=	1 ounce
16 ounces (7,000 grains)	=	1 pound
2,000 pounds	=	1 short ton
2,240 pounds	=	1 long ton

CUBIC MEASURE

2.728 cubic inches (cu. in.)	=	1 cubic foot
27 cubic feet	=	1 cubic yard

SQUARE MEASURE
144 square inches (sq. in.) = 1 square foot
9 square feet = 1 square yard

ELECTRIC

Table 17. Appliance Energy Requirements

<u>Major Appliances</u>   <u>Annual kWh</u>
Air-Conditioner (room) (Based on 1000 hours of
   operation per year. This figure will vary
   widely depending on geographic area and
   specific size of unit) 860
Clothes dryer 993
Dishwasher (including energy used to heat water)............2,100
Dishwasher only 363
Freezer (16 cu. ft.)...1,190
Freezer - frostless (16.5 cu. ft.) 1,820
Range with oven 700
   with self-cleaning oven 730
Refrigerator (12 cu. ft.) 728
Refrigerator - frostless (12 cu. ft.) 1,217
Refrigerator/Freezer (12.5 cu. ft.) 1,500
Refrigerator/Freezer - frostless (17.5 cu. ft.) 2,250
Washing Machine - automatic
   (including energy used to heat water) 2,500
Washing Machine - non-automatic
   (including energy to heat water) 2,497
   washing machine only 76
Water Heater 4,811

Kitchen Appliances
Blender 15
Broiler 100
Carving Knife 8
Coffee Maker 140
Deep Fryer 83
Egg Cooker 14
Frying Pan...186

Table 17. Appliance Energy Requirements (continued)

Major Appliances	Annual kWh
Hot Plate	90
Mixer	13
Oven, Microwave (only)	190
Roaster	205
Sandwich Grill	33
Toaster	39
Trash Compactor	50
Waffle Iron	22
Waste Disposer	30

Heating and Cooling

Air Cleaner	216
Electric Blanket	147
Dehumidifier	377
Fan (attic)	281
Fan (circulating)	43
Fan (rollaway)	138
Fan (window)	170
Heater (portable)	176
Humidifier	163

Laundry

Iron (hand)	144

Health and Beauty

Germicidal Lamp	141
Hair Dryer	14
Heat Lamp (infrared)	13
Shaver	1.8
Sun Lamp	16
Toothbrush	.5
Vibrator	2

Home Entertainment

Radio	86
Television	
Black and White	
Tube Type	350
Solid State	120

Table 17. Appliance Energy Requirements (continued)

Major Appliances	Annual kWh
Color	
Tube Type	660
Solid State	440
Housewares	
Clock	17
Floor Polisher	15
Sewing Machine	11
Vacuum Cleaner	46

Table 18. Alternating Current Calculations

To Calculate	Alternating Current Three-Phase	Alternating Current Single-Phase
Amperes when horse power is known	$\dfrac{H.P. \times 746}{1.73 \times E \times \%Eff \times P.F.}$	$\dfrac{H.P. \times 746}{E \times \%Eff \times P.F.}$
Amperes when kilowatts are known	$\dfrac{K.W. \times 1000}{1.73 \times E \times P.F.}$	$\dfrac{K.W. \times 1000}{E \times P.F.}$
Amperes when K.V.A. are known	$\dfrac{K.V.A. \times 1000}{1.73 \times E}$	$\dfrac{K.V.A. \times 1000}{E}$
Kilowatts	$\dfrac{1 \times E \times 1.73 \times P.F.}{1000}$	$\dfrac{1 \times E \times P.F.}{1000}$
K.V.A.	$\dfrac{1 \times E \times 1.73}{1000}$	$\dfrac{1 \times E}{1000}$
Horsepower (Output)	$\dfrac{1 \times E \times 1.73 \times \%Eff \times P.F.}{746}$	$\dfrac{1 \times E \times \%Eff \times P.F.}{746}$

E = Volts. K.W. = Kilowatts. P.F. = Power Factor. I = Amperes
%Eff = Percent Efficiency. K.V.A. = Kilovolt amperes.
H.P. = Horsepower

Table 19. Kilowatt Conversion Factors
Kilowatt Conversion Factors

1 kilowatt	=	1.3415 horsepower
	=	738 ft lb per sec
	=	44,268 ft lb per min
	=	2,656,100 ft lb per hr
	=	56.9 Btu per min
	=	3,413 Btu per hr

Table 20. Effects of Electric Current on Humans

Current Values	Effect
1 ma	Causes no sensation
1 to 8 ma	Sensation of shock. Not painful
8 to 15 ma	Painful shock
15 to 20 ma	Cannot let go
20 to 50 ma	Severe muscular contractions
100 to 200 ma	Ventricular fibrillation
200 & over ma	Severe burns. Severe muscular contractions

Table 21. Equivalent Electrical Units

1 Kilowatt	=	1,000 Watts
1 Kilowatt	=	1.34 H.P.
1 Kilowatt	=	44,260 Foot-Pounds per minute
1 Kilowatt	=	56.89 B.T.U. per minute
1 H.P.	=	746 Watts
1 H.P.	=	33,000 Foot-Pounds per minute
1 H.P.	=	42.41 B.T.U. per minute
1 B.T.U.	=	778 Foot-Pounds
1 B.T.U.	=	0.2930 Watt-Hour
1 Joule	=	1 Watt-Second

Table 22. United States Power Characteristics

	Voltage	Amperes	Phase
Controls	20 to 12-	5 to 15	Single
Small Equipment	120 208 240 277	10 to 40	Single
Large Equipment	208 240 480	30 to 400	Three

Table 23. Induction Motor Synchronous Speeds

Poles	@ 60 Hz
2	3,600
4	1,800
6	1,200
8	900
10	720
12	600

FORMULAE

Table 24. Geometric Formulas

Circumference of a circle	$C = \pi d$
Length of an arc	$L = \dfrac{n}{360} \times \pi d$
Area of a rectangle	$A = LW$
Area of a square	$A = S^2$

Area of a triangle	$A = 1/2 bh$
Area of a trapezoid	$A = 1/2 h(b + b')$
Area of a circle	$A = .7854 d^2$, or $1/4 \pi d^2$
Area of a sector	$S = \dfrac{n}{360} \times .7854 d^2$
Area of an ellipse	$A = .7854 ab$
Area of the surface of a rectangular solid	$S = 2LW + 2LH + 2WH$
Lateral area of a cylinder	$S = \pi dh$
Area of the surface of a sphere	$S = \pi d^2$
Volume of a rectangular solid	$V = LWH$
Volume of a cylinder	$V = .7854 d^2 h$
Volume of a sphere	$V = .5236 d^3$, or $1/6 \pi d^3$
Volume of a cube	$V = e^3$

Heat

Table 25. Heat Equivalents

1 Btu =		252 calories
1 kilocalorie	=	1000 calories
1 Btu/lb.	=	.55 kcal/kg
1 Btu/lb.	=	2.326 kj/kg
1 kcal/kg	=	1.8 Btu/lb
1 Btu/hr	=	0.2931 watts

Table 26. Heat Generated by Appliances

General lights and heating	3.4 Btu/hr/watt
2650 watt toaster	9100 Btu/hr
5000 watt toaster	19,000 Btu/hr
Hair dryer	2000 Btu/hr
Motor less than 2 HP	3600 Btu/hr/HP
Motor over 3 HP	3000 Btu/hr/HP

Table 27. Heat Loss from Hot Water Piping

Pipe Size, Inches	Hot Water, 180°F Bare	Insulated
1/2	65	22
3/4	75	25
1	95	28
1-1/4	115	33
1-1/2	130	36
2	160	42
2-1/2	185	48
3	220	53
4	280	68

Table 28. Heat Loss from Bare Steel Piping
(Btu/hr @ 70 deg F ambient)

	Hot Water		Steam	
Nominal Pipe Size	120F	180F	5 psi	100 psi
1/2	0.455	0.546	0.612	0.760
1	0.684	0.819	0.919	1.147
2	1.180	1.412	1.578	1.987
4	2.118	2.534	2.850	3.590
8	3.880	4.638	5.210	6.610
12	5.590	6.670	7.500	9.530

Table 29. Specific Heats of Common Substances

Aluminum	.2143
Brine	.9400
Coal	.314
Copper	.0951
Ice	.5040
Petroleum	.5110
Water	1.000
Wood	.3270

Table 30. Heat Content of Common Fuels

Number 6 fuel oil	152,400 Btu per gallon
Number 2 fuel oil	139,600 Btu per gallon
Natural gas	950 to 1150 Btu per cubic foot
Propane	91,500 Btu per gallon

Table 31. Typical Fuel Oil Analysis

	California Hi Sulphur/Lo Sulphur		Texas Hi Sulphur/Lo Sulphur	
% Sulphur	1.0	/4.2	1.0	/2.8
% Total Ash	0.1	/0.08	0.08	/0.10
% Carbon	87.2	/85.2	86.8	/86.1
% Hydrogen	10.0	/10.0	10.8	/10.1
% Oxygen/Nitrogen	1.8	/1.0	1.0	/1.0
Moisture	.03	/		/0.2
Specific Gravity (@ 60F)	1.007	/0.986	0.977	/1.003

Miscellaneous

Table 32. Cylindrical Tank Capacity

Capacity (gal)	Diameter (in)	Length	
100	24	4'	
150	30	4'	
250	30	7'	
500	42	7'	
550	48	6'	
750	48	8'	
1000	65	6'	
1500	48	16'	
2000	65	11'	11.5"
3000	65	17'	10"
4000	65	23'	8.5"
5000	72	23'	9"
10000	108	21'	

Table 33. Metric Capacity

Name		Capacity
Milliliter (ml.)	=	.001
Centiliter (cl.)	=	.01
Liter (l)	=	1.
Decaliter (Dl.)	=	10
Hectoliter (Hl.)	=	100
Kiloliter (Kl.)	=	1,000
Myrialiter (Ml.)	=	10,000

Table 34. Metric Length

		METERS
Millimeter (mm.)	=	.001
Centimeter (cm.)	=	.01
Decimeter (dm.)	=	.1
Meter (m)	=	1
Decameter (Dm.)	=	10
Hectometer (Hm.)	=	100
Kilometer (Km.)	=	1,000
Myriameter (Mm.)	=	10,000

Table 35. Metric Weight

Name	Grams
Milligram (mg.)	.001
Centigram (cg.)	.01
Decigram (dg.)	.1
Gram (g)	1
Decagram (Dg.)	10
Hectogram (Hg.)	100
Kilogram (Kg.)	1000
Myriagram (Mg.)	10,000
Quintal (Q.)	100,000
Tonneau (T.)	1,000,000

Table 36. Weights of Metals

Name of Metal	Pounds/Cu. Ft.
Aluminum	166
Brass	504
Copper	550
Iron	450
Lead	712
Silver	655
Steel	490
Tin	458
Zinc	437

Table 37. Ultimate Bending Strength

Material	PSI	Material	PSI
Cast iron	30,000	Hemlock	3,500
Wrought iron	45,000	Oak, white	6,000
Steel	65,000	Pine, white	4,000
Stone	1,200	Pine, yellow	7,000
Concrete	700	Spruce	3,000
Ash	8,000	Chestnut	4,500

Table 38. Recommended Lighting Levels

Area	Foot-Candles
Perimeter of building	5
Office areas	70
Corridors, elevators and stairways	20
Toilets and washrooms	30
Entrance lobbies	10
Dining areas	20
Mechanical rooms	20

STEAM

Table 39. Steam Pressure Temperature Relationship

GAGE PSI	SAT TEMP F	GAGE PSI	SAT TEMP F
0	212	70	316
5	228	80	324
10	240	90	331
20	259	100	338
30	274	200	388
40	287	300	422
50	298	400	448
60	308		

Table 40. Steam Trap Selection

Characteristic	Inverted Bucket	Thermostatic
Method of Operation	Intermittent	Intermittent
Steam Loss	None	None
Resistance to Wear	Excellent	Good
Corrosion Resistance	Excellent	Excellent
Resistance to Hydraulic Shock	Excellent	Poor
Vents Air and Co-2 at Steam Temperature	Yes	No
Ability to Vent Air at Very Low Pressure	Poor	Excellent
Ability to Handle Start-up Air Loads	Fair	Excellent
Operation Against Back Pressure	Excellent	Excellent
Resistance to Damage from Freezing	Poor	Excellent
Ability to Purge System	Excellent	Good
Ability to Operate on Very Light Loads	Good	Excellent
Responsiveness to Slugs of Condensate	Immediate	Delayed
Ability to Handle Dirt	Excellent	Fair
Comparative Physical Size	Large	Small

TEMPERATURE

Table 41. Color Scale of Temperature

Color	Temperature	Color	Temperature
Incipient red heat	900-1100	Yellowish red heat	1800-2200
Dark red heat	1100-1500	Incipient white heat	2200-2600
Bright red heat	1500-1800	White heat	2600-2900

Table 42. Centigrade/Fahrenheit Scale

°C	°F	°C	°F	°C	°F
–50	–58	30	86	100	212
–40	–40	40	104	110	230
–30	–22	50	122	120	248
–20	–4	60	140	130	266
–10	14	70	158	140	284
0	32	80	176	150	302
10	50	90	194	160	320
20	68				

Table 43. Melting Point of Common Substances

Metal	Symbol	Degrees F
Aluminum	Al	1218
Copper	Cu	1981
Iron	Fe	2795
Lead	Pb	621
Mercury	Hg	–38
Molybdenum	Mo	4750
Silicon	Si	2590
Silver	Ag	1761
Tin	Sn	449
Tungsten	W	6100
Zinc	Zn	787

Table 44. Temperature of Waste Heat Gases

Source of Gas	Temperature, Deg. F
Ammonia oxidation process	1,350 - 1,475
Annealing furnace	1,100 - 2,000
Black liquor recovery furnace	1,800 - 2,000
Cement kiln (dry process)	1,150 - 1,350

Appendix E — Tables

Table 44. Temperature of Waste Heat Gases (continued)

Source of Gas	Temperature, Deg. F
Cement kiln (wet process)	800 - 1,100
Coke over	
beehive	1,950 - 2,300
by-produce	up to 750
Copper refining furnace	2,700 - 2,800
Copper reverberatory furnace	2,000 - 2,500
Diesel engine exhaust	550 - 1,200
Forge and billet heating furnace	1,700 - 2,200
Garbage incinerator	1,550 - 2,000
Gas benches	1,050 - 1,150
Glass tanks	800 - 1,000
Heating furnace	1,700 - 1,900
Malleable iron air furnace	2,600
Nickel refining furnace	2,500 - 3,000
Petroleum refinery still	1,000 - 1,100
Steel furnace, open health	
oil, tar, or natural gas	800 - 1,100
producer gas-fired	1,200 - 1,300
Sulfur, ore processing	1,600 - 1,900
Zinc fuming furnace	1,800 - 2,000

WATER

Table 45. Water Equivalents

U.S. Gallons	x	8.33	=	Pounds
U.S. Gallons	x	0.13368	=	Cu. Ft.
U.S. Gallons	x	231.	=	Cu. Ins.
U.S. Gallons	x	3.78	=	Litres
Cu. Ins. of Water (39.2°)	x	0.03613	=	Pounds

Table 45. Water Equivalents (continued)

Cu. Ins. of Water (39.2°)	x	0.004329	=	U.S. Gals.
Cu. Ins. of Water (39.2°)	x	0.576384	=	Ounces
Cu. Ft. of Water (39.2°)	x	62.427	=	Pounds
Cu. Ft. of Water (39.2°)	x	0.028	=	Tons
Pounds of Water	x	27.72	=	Cu. Ins.
Pounds of Water	x	0.01602	=	Cu. Ft.
Pounds of Water	x	0.12	=	U.S. Gals.

Table 46. Supply Line Sizes for Common Fixtures

Laundry Tubs	1/2 inch
Drinking Fountains	3/8 inch
Showers	1/2 inch
Water-Closet Tanks	3/8 inch
Water-Closets (with flush valves)	1 inch
Kitchen Sinks	1/2 inch
Commercial Type Restaurant Scullery Sinks	1/2 inch

Table 47. Pressure of Water

One foot of water = 0.4335 psi
One psi = 2.31 foot of water

Feet Head	Pressure Psi	Feet Head	Pressure Psi
10	4.33	55	23.82
15	6.49	60	25.99
20	8.66	70	30.32
25	10.82	80	34.65
30	12.99	90	38.98
35	15.16	100	43.31
40	17.32	200	86.63
45	19.49	300	129.95
50	21.65	400	173.27

Table 48. Water Requirements of Common Fixtures

Fixture	Cold, GPM	Hot, GPM
Water-closet flush valve	45	0
Water-closet flush tank	10	0
Urinals, flush valve	30	0
Urinals, flush tank	10	0
Lavatories	3	3
Shower, 4-in. head	3	3
Shower, 6-in. head and larger	6	6
Baths, tub	5	5
Kitchen sink	4	4
Pantry sink	2	2
Slop Sinks	6	6

Table 49. Blowdown Flow Rates - 3" Pipe

PSI	GPS	PSI	GPS
15	0.50	60	1.06
20	0.60	70	1.13
30	0.73	80	1.20
40	0.86	90	1.30
50	0.96	100	1.36

Table 50. Tank Capacities per Foot of Depth

Diameter in Feet	Gallons	Diameter in Feet	Gallons
1	5.84	9	475.00
2	23.43	10	587.00
3	52.75	12	845.00
4	93.80	14	1,150.00
5	146.80	16	1,502.00
6	211.00	18	1,905.00
7	287.00	20	2,343.00
8	376.00		

Table 51. Efficiency Loss Due to Scale

Thickness in Inches	Percent Loss
1/64	4
1/16	11
1/8	18
3/16	27
1/4	38
3/8	48
1/2	60

INDEX

accident investigation 50
acoustics 132
adjacencies 6, 26
aesthetics 20
air pollution 10
air velocities 28
analysis 14
ANSI 18
appraisal 3
as built drawings 2, 14
asbestos 14, 112, 153
assessment 10
autonomous power 77, 93
auxiliary mechanicals 33
barrier free design 9
base line information 20
benefits 6
blueprint index 156
budget justification 6
budgets 8, 167
building characteristics 7
building codes 17, 35, 134
building components 6, 75
building facades 133
building structure 130
building layout 25
CADD 21
capital outlay 5
circuit zoning 136
clean air regulation 115, 116

code violations 26, 35
combustion analyzer 115
commissioning 1
company policy 50
compartmentalization 7, 131
contingency plan 31, 148, 149, 155
contract services 169
controlling strategies 44, 45
cooling tower 30
coordination of effort 25
corrective action 45
corrective action plans 125, 175, 182, 184, 185
cross connections 31
data collection 20, 21
debriefing 21
department operations 8
design loading 7
dielectric strength 91
documenting 20
duct penetrations 76
eddy current testing 19
electric shock 81, 82
electrical circuits 84, 85
electrical current 81, 82
electrical hazards 198
electrical inspection 86-90
electrical maintenance 84
electrical protective gear 85
electrical safety 81-103

electrical systems 14
emergency generator 14, 145
emergency power 93
employee orientation 62
employee training 165
energy efficiency 138
engineered exhaust 77
envelope of protection 85
environmental compliance 105, 152-154
environmental concerns 57
environmental factors 39
equipment age 10, 13
equipment hazards 199
equipment loading 31
ergonomics 6, 20
evaluation
    diagnostic 2-5
    indicative 2-4
    investigative 2-4
evaluation format 13, 16
evaluation planning 13
evaluation report 175-180
exits 77
extent of need 25
extinguishing systems 78
feedback 20
fire and safety 42, 46
fire brigades 70, 78
fire codes 69
fire extinguishers 94
fire hazards 200
fire protection 29, 69, 150-152
fire resistant materials 76
fire safety program 71-74
flammable storage 79, 80
floor assembly 76
floor loading 132
forms/schedules 19, 158
freeze protection 142
full load tests 19
functional adjacencies 131

functional elements 6, 20
generator testing 93, 95-103
ground fault interruption 91, 92
grounding 92
grounds 32
grounds maintenance 128, 180
handicap access 7
hazardous waste 114, 153
hierarchy charts 18
high voltage substations 84
hours of operation 10
human factors 57
ID badges 21
impairment potential 37
incident reporting 50
incidents/accidents 168
ingress/egress 126
inspection 2
instrumentation 19
insurance 38
integrity judgements 18
interior finishes 20, 33
interstitial spaces 132
inventories 8
jurisdictional compliance 163
kick off meeting 21
labor and material 6
labor relations 16
landscaping 129
licenses/permits 17, 139, 162
life expectancy 91
life safety code 75, 77
load bearing walls 7, 130
locked rotor values 93
machinery histories 168
maintenance costs 10
master lists 18
mechanical components 88
mechanical systems 8
meeting quarters 19
micro climates 7

Index

moveable partitioning 35
MSDS 108, 109
NEC 17, 91
NFPA 17, 69, 75
nuisance tripping 92
operating costs 10
operating history 25
operating logs 165
operating manuals 10
operating procedures 46
orientation 7, 26
OSHA 85, 110
panelboards 86
PCB's 14, 123
peak shaving 93
penthouse 33, 132
peril identification 37
pesticides 121, 122
petroleum storage 106, 107
photo cells 32
physical elements 6
physical separation 76
POE 2-4, 19, 125, 175
policies/procedures 8, 39, 49, 55, 57, 70, 94, 95-103, 157, 160
post occupancy 2
power house 30
power plant 40
power/lighting 19, 28, 32
preventive maintenance 90, 134
prioritizing 183
procedural factors 39
procedural problems 57
program characteristics 49
program compliance 201
program components 50
program evaluation 50
project close out 1, 2
project manager 2
project planning 158
punch list 2

pure risk 38
quality assurance 10
questionnaires 20, 21, 26
recommendations 181
redundancy 31, 136, 137, 140
reference materials 18, 20
repairs 19
report layout 175
reporting 14, 16, 19
reports 167
review documents 17
regulatory compliance 10
right-to-know 62
risk 37-47
risk assessment 41
risk exposure 39, 125
risk factors 125
risk management 43
risk manipulation 43
risk parameters 41, 42
risk valuations 125
rounds/routines 39
safe drinking water 120
safety 27
safety charter 55, 187-195
safety committee 55
safety department 50
safety hazard worksheets 197-204
safety interlocks 140
safety management 49-67
safety programs 49, 84, 147
safety rules 62, 164
safety tips 93, 94
sampling 21
scheduling shut downs 21
security 202
sensor time delays 92
service agreements 19
service interruptions 21
service manuals 2
single phase protection 92, 93

site characteristics 7
smoke compartment 76
smoke dampers 76
SOP manual 14, 157
sources of risk 39
specifications 2, 14, 18
speculative risk 38
stack emissions 10, 32, 46
staffing 16, 55, 159
staging area 19, 20
statement of construction 17, 70, 76, 205-221
stationary engineering 30
storage areas 79, 80
storage tanks 78
structural containment 76, 77
structural failures 16
structural hazards 203
supporting documentation 182
surveys 10, 19, 158
system commissioning 1
system components 90
system routes 25
systems assessment 19
systems maintenance 49, 90
team leader 25
technical library 50, 156
temporary solutions 16

tenant interviews 26
testing & calibration 43
thermal overload 93
thermographic studies 19, 91
time requirements 18, 26
timers 32
tools 27
touring 29
toxic clean up 123
training 160
transformers 86, 93
trends 20
unbalanced voltages 93
underground tanks 14, 19, 32
unit capacity 31
unit concept 76
utilities 29, 166
valve/switch lists 136
volatile liquids 80
walk through inspection 25
water quality 117, 119
water treatment 19, 22, 23, 29
ways of travel 77
work clearances 85
work requisitioning 11, 158
workers right-to-know 110
zoning 25